Disaster and Risk Research: GADRI Book Series

Disaster and Risk Research: GADRI Book Series is published under the auspices of the Global Alliance of Disaster Research Institutes (GADRI). The global status of disaster research reflects the major strides made in the disaster sciences. These volumes present the forefront of disaster research, including key scientific findings, methodologies, policy recommendations and case studies. Whilst disaster risk needs to be managed in an integrated manner, persistently isolated applications of knowledge, practice and policy are falling short of ensuring disaster-resilient societies. Responding to this deficit calls for measurement, tools, techniques and institutional structures that can realistically support comprehensive risk assessment and management across multiple hazard landscape. As such, disaster research is now faced with a multi-disciplinary, multi-stakeholder challenge. Contributions to this series therefore address many varied and critical opportunities to advance the subject area. A cross-cutting vision shared across the Disaster and Risk Research volumes is to improve the future of scientific and technological guidance with clearly identifiable roadmaps to ensure human safety and security. The Global Alliance of Disaster Research Institutes was established in March 2015, directly after the United Nations World Conference on Disaster Risk Reduction (WCDRR 2015) in Sendai, Japan, based on the belief that a multi-institutional alliance can strengthen disaster research and its influences around the world. GADRI has a mandate to support the implementation of the Sendai Framework for Disaster Risk Reduction 2015-2030 and is a member of the Scientific and Technical Advisory Group (STAG) of the United Nations Office for Disaster Risk Reduction (UNISDR). In addition, GADRI provides a platform for scientific communities from different disciplines, backgrounds and countries, helping them share their knowledge, findings and views. This approach yields more holistic and farther-reaching insights, which can contribute to further steps in effective disaster risk management.

More information about this series at http://www.springer.com/series/16177

Hirokazu Tatano · Andrew Collins
Editors

Proceedings of the 3rd Global Summit of Research Institutes for Disaster Risk Reduction

Springer

Editors
Hirokazu Tatano
Disaster Prevention Research Institute
Kyoto University
Uji, Kyoto, Japan

Andrew Collins
Department of Geography
and Environmental Sciences
Northumbria University
Newcastle upon Tyne, UK

ISSN 2524-5961 ISSN 2524-597X (electronic)
Disaster and Risk Research: GADRI Book Series
ISBN 978-981-15-8661-3 ISBN 978-981-15-8662-0 (eBook)
https://doi.org/10.1007/978-981-15-8662-0

This Springer imprint is published by the registered company Springer Nature Singapore Pte Ltd.
The registered company address is: 152 Beach Road, #21-01/04 Gateway East, Singapore 189721, Singapore

Preface

This book is a compendium of the keynote speeches and the oral and poster presentations of the 3rd Global Summit of Research Institutes for Disaster Risk Reduction: Expanding the Platform for Bridging Science and Policy.

Speakers and presenters were specifically requested to focus on the sub-theme of the conference and share activities and experiences in bridging gaps in research, science and policy areas for disaster risk reduction and efforts to stream in science into decision-making processes.

The following highlights a few of the outcomes of the 3rd Global Summit.

The 3rd Global Summit was held at the Disaster Prevention Research Institute (DPRI), Kyoto University in Kyoto, Japan, from 19 to 21 March 2017. It was attended by members representing institutions that form the Global Alliance of Disaster Research Institutes (GADRI) together with other participants from around the world.

By acknowledging the need for efforts, continuity and improvement among the international community in disaster risk reduction, GADRI is committed to

- Coordinate the smooth cooperation among GADRI member research institutions.
- Increase efforts to engage, especially young researchers and scientists, as well as practitioners, policy-makers, educators and media in various fields of disaster risk reduction.
- Further knowledge, priorities and directions effectively in the field of disaster risk reduction.
- Engage networking towards improving communication among members and other entities in the direction of collaborative international research, enhance timely promotion of research results, open communication to share data and other research opportunities. It was noted that integration and collection of big data sets and its usage through advanced database systems are an integral part of research in disaster risk reduction.

- Prioritize dissemination of research results related to GADRI activities in the area of disaster risk reduction; GADRI will enhance its efforts to manage knowledge transfer through books published under the GADRI Book Series, research journals, via online communication and other methods of educational exchange and collaborative opportunities.
- Promote the development of research capacity; GADRI will enhance its efforts in the area of disaster risk reduction and encourage in particular a wide range of young scientists and researchers to actively engage in the field.
- Improve its support of the Sendai Framework Agenda for 2015–2030 and the Science and Technology Roadmap, as mandated in the Charter of GADRI. During the GADRI 3rd Global Summit, the Sendai Framework Priority Areas of "Understanding, Governance, Resilience and Recovery" were discussed as core topics through group discussion sessions covering a broad area of disciplines to identify gaps in disaster risk reduction research. GADRI will continue to encourage the scientific community to bolster its efforts in the areas of research, implementation, education and policy-making and to effectively and actively contribute towards the Sendai Framework Agenda.

The Disaster Prevention Research Institute (DPRI), Kyoto University, and other sponsors were applauded by the participants for hosting the 3rd Global Summit successfully. Participants also expressed their appreciation for hosting the GADRI Secretariat at DPRI, Kyoto University, Japan.

Prof. Hirokazu Tatano
Secretary-General
Global Alliance of Disaster Research Institutes (GADRI)

Professor
Disaster Prevention Research Institutes (DPRI)
Kyoto University
Uji, Japan
e-mail: tatano.hirokazu.7s@kyoto-u.ac.jp

Prof. Andrew Collins
Chair
GADRI Board of Directors

Leader
Disaster Development Network (DDN)
Northumbria University
Newcastle upon Tyne, UK
e-mail: andrew.collins@northumbria.ac.uk

About This Book

This book presents selected papers from the 3rd Global Summit of Research Institutes for Disaster Risk Reduction which focused on "Expanding the Platform for Bridging Science and Policy Making". The conference not only provided a platform for discussion and exchange of information on most important current and future research projects in disaster risk reduction and management but also promoted active dialogue through group discussion sessions that were held according to the disaster research disciplines. To facilitate group discussion sessions, a prior survey was conducted to evaluate the current research status and identify the most important future research themes and projects.

The book will benefit a broad readership of researchers, practitioners, policy-makers, educators and students.

Contents

Part I
Overview of the 3rd Global Summit of Research Institutes for Disaster Risk Reduction

Chapter 1
Summary Report of the 3rd Global Summit of Research Institutes for Disaster Risk Reduction: Expanding the Platform for Bridging Science and Policy Making (GSRIDRR 2017) Held at the Disaster Prevention Research Institute (DPRI), Kyoto University, Kyoto, Japan, from 19th to 21st of March 2017

Andrew Collins, Hirokazu Tatano, and Wilma James

Abstract The Sendai Framework for Disaster Risk Reduction 2015–2030 draws our attention to work together towards reducing damages caused disasters to livelihoods and economies, especially in developing countries by 2030. This paper provides a brief summary of the 3rd Global Summit of Research Institutes for Disaster Risk Reduction under the theme of Expanding the Platform for Bridging Science and Policy Making (GSRIDRR 2017) successfully organized by the Global Alliance of Disaster Research Institutes (GADRI), and the Disaster Prevention Research Institutes (DPRI), Kyoto University at DPRI, Kyoto University, Uji Campus, Kyoto, Japan, from 19th to 21st March 2017. It reflects upon the keynote speeches, group discussion session and other activities that took place during the summit.

This chapter is a revised version of GADRI ACTIONS (2017) https://gadri.net/resources/pdf/GANewsletter_3_April2017.pdf

---

A. Collins
Disaster Development Network (DDN), Northumbria University, Newcastle, UK
e-mail: andrew.collins@northumbria.ac.uk

H. Tatano · W. James (✉)
Global Alliance of Disaster Research Institutes (GADRI), Kyoto University, Kyoto, Japan
e-mail: james.wilmatheonesta.6r@kyoto-u.ac.jp

H. Tatano
e-mail: tatano.hirokazu.7s@kyoto-u.ac.jp

Disaster Prevention Research Institutes (DPRI), Kyoto University, Kyoto, Japan

H. Tatano and A. Collins (eds.), *Proceedings of the 3rd Global Summit of Research Institutes for Disaster Risk Reduction*, Disaster and Risk Research: GADRI Book Series,
https://doi.org/10.1007/978-981-15-8662-0_1

Keywords 3rd Global Summit · Research institutes · Disaster risk reduction · Bridging science and policy making · Disasters · Geohazards · Hazards · Earthquakes · Tsunamis · Flash floods

The Opening Ceremony of the conference was attended by Prof. Kaoru Takara, Deputy Executive Director, Kyoto University; and Director, Disaster Prevention Research Institute (DPRI), Kyoto University; Mr. Shuichi Yamauchi, Vice-Governor of the Kyoto Prefecture and Mr. Tadashi Yamamoto, the Mayor of Uji City.

Prof. Hirokazu Tatano, Secretary-General, GADRI

The establishment of GADRI dates back to November 2011 when the Disaster Prevention Research Institute (DPRI), Kyoto University, Japan, organized First Global Summit during the aftermath of 2011 triple disaster, Great Tohoku Earthquake, tsunami and the nuclear disaster. Resolution adopted during the First Global Summit proposed an international alliance of research institutes involved in disaster risk reduction and risk management fostered by DPRI, Kyoto University, Kyoto, Japan.

A Second Global Summit of Research Institutes for Disaster Risk Reduction was organized by DPRI, Kyoto University, Kyoto, Japan, in March 2015 soon after the Third UN World Conference on Disaster Risk Reduction was held in Sendai, Japan. The purpose of the Second Global Summit was to take stock of progress on initiatives recommended at the First Global Summit. As a final outcome of the conference, the Global Alliance of Disaster Research Institutes (GADRI) was established. GADRI supports the implementation of the Sendai Framework for Disaster Risk Reduction 2015 to 2030 (SFDRR) and the work of the Scientific and Technical Advisory Group of the United Nations Office for Disaster Risk Reduction (UNISDR). Currently, GADRI membership stands around 130 institutes.

The Third Global Summit brought together GADRI members and other participants totalling an attendance of 251 members from 102 institutions in 38 states among which were UN and governmental agencies, international organizations, private section, academia and other areas and disciplines involved in disaster risk management and prevention.

Starting with the opening ceremony, the conference moved on to feature 11 keynote speakers from the UN, Japanese Government, academia and others, various group discussions and plenary sessions. Sufficient time was allocation for oral and poster presentations too. Third Global Summit closed with a wrap-up session, field visits and a closing banquet which was complimented by a cultural performance from the Kyoto Prefecture.

GADRI Secretariat appreciated and valued the support and participation by its member institutes.

1.1 Third Global Summit

The Third Global Summit focused focusing on the four Priority Areas of the Sendai Framework for Action 2015–2030, targeted an audience from the academia, international organizations, private sector and governmental agencies directly related in the area of disaster risk reduction and research with the following objectives:

- To serve as an advocate for key research policy statements that are in line with real, evidence-based disaster research needs.
- To carry out a more detailed assessment of key research challenges and to identify priority research areas.
- To identify pioneering scientific initiatives to effectively reduce the gaps between science and practice in disaster risk reduction activities.
- To share and build on achievements, and outcomes of past and ongoing GADRI activities addressing research gaps; and
- To foster links between local and international organizations and their programmes through the GADRI network.

Foreword from Kyoto University President, Prof. Juichi Yamagiwa was read by Prof. Kaoru Takara, Deputy Executive Director, Kyoto University; and Director, Disaster Prevention Research Institute (DPRI), Kyoto University. Prof. Yamagiwa stressed the importance of alliance in embarking upon disaster risk reduction and spreading the word on raising awareness across borders to achieve optimum results in disaster risk reduction and resilience to disasters.

While welcoming everyone to Uji City, Mr. Tadashi Yamamoto, Mayor, Uji City, Kyoto, Japan, extended his congratulations to organizers of the Third Global Summit. He also expressed his deep appreciation to the Disaster Prevention Research Institute (DPRI), Kyoto University for their continued support to the Uji City and appreciated the work conducted by DPRI and all other participating institutes on the area of disaster risk reduction and their continuous use of results to support areas affected by disasters and the effective implementation of administrative disaster prevention policies by governments.

From left: Prof. Andrew Collins, Prof. Kaoru Takara, Uji City Mayor Tadashi Yamamoto and Kyoto Prefecture Vice-Governor Shuichi Yamauchi

In the welcome message delivered by the Vice-Governor of Kyoto City, Mr. Shuichi Yamauchi, he stressed that Japan is prone to suffer disasters more than any other country with many earthquakes, torrential rains and other natural disasters. The 2011 Great Eastern Japan Earthquake caused enormous damage to the region. After six years, recovery operation is still underway. Considering the fact of large-scale disasters frequently occurring around the world, Third Global Summit is paving a foundation to contribute promotion of disaster risk reduction and disaster countermeasures worldwide. This in turn will encourage other research programmes on disaster prevention and mitigation to join together to build safe and resilient societies.

Mr. Tadashi Yamamoto, Mayor of Uji City, referred to the heavy rain fall in 2012 southern parts of Kyoto where more than 2000 buildings were damaged and roads, agriculture, forestry, tea planting and rice fields inundated by floods and landslides. He referred to the invaluable assistance received from DPRI in assisting with the disaster prevention plans. The summit, he believed, is a positive forum to share research results through various interchanges with people from all over the world to deepen disaster prevention research and policies.

Each day of the conference started with an impressive array of keynote speakers who shared their experiences and shed light on policy issues and stimulated a platform for dialogue between various stakeholders.

1.1.1 *Theme Day One—19 March 2017: Connecting with the International Community and Initiatives for Collaborative Activities: Listening to the Opinion of the International Organizations and Other Stakeholders*

First-day keynote speeches by a few of the UN agencies and the Cabinet Office of the Japanese Government focussed on contributions to reduce disasters and increase community awareness programmes. They shared their experiences and view points where more efforts could be invested in the future on collaborative activities.

Ms. Setsuko Saya, Director, Disaster Management Bureau, the Cabinet Office of the Government of Japan

Ms. Setsuko Saya, Director, Disaster Management Bureau, the Cabinet Office of the Government of Japan, is also the Head of the Japanese Delegation to UNDRR.

Her topic was on "Collaboration between Science and Policy Making in Japan". She shared information on best practices used in Japan to communicate with the science and technology community and how the Government of Japan integrates these inputs into policy making processes. Science and technology plays an integral part in the policy and decision makings of the Government of Japan.

Lessons learned the Kumamoto earthquakes were highlighted with examples. She stated that situation was quite unusual due to its nature of occurring in a less earthquake-prone area in Japan, which required much effort by the national and local governments for emergency response when evacuating 180,000 people.

The Government of Japan considers the collaboration with the science and technology community as indispensable for policy makers. Many governments have access to raw data and their own governmental methodology to contribute to disaster risk reduction. Provision of data is somehow a mechanical procedure. The problem is how to address this problem and how to interpret the data for policy making processes.

This keynote speech concluded with two proposals on how S&T could communicate to policy makers and communities:

- by augmenting S&T in educational systems; and
- by support to assessing indicators proposed by UNDRR (UNDRR proposed 38 indicators of disaster risk reduction).

The UNISDR was represented by Dr. Chadia Wannous, Senior Advisor, United Nations Office for Disaster Risk Reduction, Geneva, Switzerland. In her keynote speech on "The role of Science and Technology in the Implementation of the Sendai Framework 2015–2030", she shared information on various efforts in place to manage future risks and reduce disaster risk. In addition, national and regional DRR platforms, partners and networks are planning to integrate S&T actors in their projects and functionalities. She called to GADRI to use its position among the global stakeholders to further facilitate the contributions from the science and technology community for the implementation of the Sendai Framework Agenda. In concluding, she stated that preparedness and resilience in social, economic and environmental aspects are necessary and called for comprehensive and a structured contribution from the science and technology community and further reiterated the need to collaborate for action.

Prof. Virginia Murray, Vice-Chair, Scientific and Technical Advisory Group (STAG), UNISDR, delivered a keynote speech on "Science and Technology Commitment to the Implementation of the Sendai Framework for Disaster Risk Reduction 2015–2030" where she discussed science and policy linkages at national levels. She emphasized the need for a formal national disaster risk reduction science policy council/platform or a form of national focal points for science to support DRR and management plans to be in place to achieve these goals. Conducting periodic review of knowledge needs, new science (including implementation science) and identification of research gaps are some tools that could be used for the purpose. More effort is needed to work out how to achieve this and ensure avoiding duplication of efforts. She shared some reflections and recommendations on the incredible work of the S&T community: This highlighted the importance of having an effective process for sharing knowledge and building partnerships. There was a need for clarity on outputs that would benefit the S&T programme for the Sendai Framework and a joined-up approach to implement its goals.

Dr. Yasusuke Tsukagoshi, Special Representative, The World Bank Office in Tokyo, discussed about "Mainstreaming Disaster Risk Management in the World Bank: Sharing Knowledge and Investing to Enhance Resilience in Developing Countries". The World Bank addresses key challenges related to disaster risk reduction in science and policy. Early warning systems combining observation and evacuation systems are tools that can save many lives. Although these constitute a crucial "missing link" for disaster resilience in many countries, disaster risk financing strategies are helping countries build layers of protection against disasters. However, to implement good governance, it is important to induce collection of data, scientific analysis and evidence. Governance is fundamental for disaster risk management and governments in each country should strengthen their policies.

1.1.2 Theme Day 2—20 March 2017—Discussions on the Future Directions of Disaster Risk Reduction, 20 March 2017

Second-day theme on future directions for disaster risk reduction discussed existing challenges along the lines of climate change and sustainable development goals which are interrelated with disaster risk reduction and resilience. The session directed towards the dimensions of evaluating indexes for future disaster risk initiatives and identifying current research gaps. A key question was how to expand activities with a renewed sense of perseverance to prevent future disaster risks by enhancing collaboration at the global scale so as to realize the targets/goals set out in the Sendai Framework in the real world.

Prof. Shahbaz Khan, Director, UNESCO Jakarta Office, Regional Science Bureau for Asia and the Pacific, focussed on "Science Support for Disaster Risk Reduction". He shared information on some of the projects undertaken by UNESCO to assist countries strengthen their capacities in disaster risk management and climate change issues especially in the area of education and through raising awareness in a culturally sensitive manner. To accomplish the goals of the Sendai Framework Agenda, he pointed out that GADRI could contribute to the efforts are most needed through international cooperation and bridging the gap between the scientists, practitioners, policy makers and the community.

Prof. Shahbaz Khan, Director, UNESCO, Jakarta Office

Prof. Kimio Takeya, Distinguished Technical Advisor to the President of Japan International Cooperation Agency (JICA) and Visiting Professor, Tohoku University, discussed the "Donor's practical point of view 'what we expect from Science and Technology Group' in relation to the Sendai Framework". As the world's financial biggest donor for DRR, JICA and Prof. Takeya has significantly contributed to formulate the Sendai Framework Document as a part of the substance of a main negotiator for the Japanese Government Team from a donor's point of view. Currently, Japan provides 64% of the financial support for disaster risk reduction and is the number one donor in the world. This in turn is Japan's contribution to the Sendai Framework. The Sendai initiatives provide knowledge and guidelines to practitioners in the governments and how to enhance political and social research.

Dr. Gabriel Duque, Ambassador of Colombia to Japan, Embassy of Colombia, Tokyo, delivered a speech on "Research on Management of Disasters and Risk Reduction: The Colombian Case" which focussed on the Colombian experience and approximation to Sendai Framework. He stressed that international cooperation in disaster risk management are important tools for developing countries to set up structures by linking research to policy making and in disaster management, and to adopt a political commitment to include in their national risk management plans.

1.1.3 Theme Day Three—21 March 2017—Sharing Information of Research Institutions, 21 March 2017

This session looked in to current research and knowledge gaps, and the status of sharing of information. Speakers from various backgrounds including practitioners from government, national and academia shared much needed and valued information on tools, resources and portals to share knowledge on disaster risk reduction. The session paved an opportunity for participants to learn about many efforts that are already in place to prevent new and reduce existing disaster risks. It was noted that networks such as GADRI could play a mega role in maximizing efforts to share and collaborate across institutions, and policy makers, and contribute to goals set out by various other stakeholders.

Dr. Tom De Groeve, Acting Head of Unit, Disaster Risk Management Unit, European Commission, Joint Research Centre (EC-JRC), delivered an informative presentation on the European Commission and the Joint Research Centre achievements in science and technology. He stressed that without the norms of communication or bridges among scientists and policy makers, DRR efforts cannot materialize. The Sendai Framework possesses a good interface with science and the EU works with the Sendai Framework and the legislation within the union for protection mechanisms for disaster management to improve the knowledge base. Science is not optimally used at the moment as networks are fragmented. Nevertheless, it does not mean that one system would fit all governments. In order to accomplish the targets, it means coming out of the comfort zones and making a contribution that would help everyone. He concluded his message by stating that networks such as GADRI are better positioned to build on partnerships, identifying research knowledge gaps and contributing as a whole to the global community.

Dr. Srikantha Herath, Senior Advisor, Sri Lanka Ministry of Megapolis and Western Development and Visiting Professor at the United Nations University, Tokyo, discussed "Knowledge to Sustainable Practices: The International Network for Transdisciplinary Education (INATE) Experience" citing experiences and case studies. This asked how to translate knowledge to sustainable practices to achieve sustainability? Higher education championed translating knowledge in to practices. He stated that one of the difficulties in adaptation is when one needs to adapt to a future which is uncertain, and resilience is much more of a pragmatic approach when trying to modify existing programmes by introducing new elements to improve the resilience of the system. When the future is uncertain, it is difficult to define future targets and objectives. Sustainability provides the framework on which sustainable goals are agreed upon. Who translates knowledge into practices? The focus here was on higher education and how they could be instrumental in translating this knowledge in to sustainable practices. Implementing programmes for longer term use requires a transdisciplinary approach. In conclusion, to adapt to challenges of rapid global change, it is easier to take a transdisciplinary approach where all actors and stakeholders could play a role to facilitate this process; and GADRI could encourage higher education which can play a major role in bringing up these transformations.

Prof. Toshio Koike, Director, International Centre for Water Hazard and Risk Management (ICHARM), discussed conventional and modern systems of disaster risk reduction and methodology in terms of increasing water-related disasters and how best to address these problems in a holistic manner. Various projects on flood and drought monitoring management and various countermeasures are in place. Seasonal prediction of agricultural drought provides information used for early warning systems. Future climate predicts serious seasons of drought with grave economic impacts. In summary, by accumulating data on damage, hazard and socio-economic processes, the data could be integrated to risk assessment. Change of identification requires monitoring and prediction. Once this information is in place, it could support policy makers and communities. He concluded by emphasizing the need for strong capacity building programmes in higher education to build up leaders and practitioners and that GADRI should take part in these efforts.

1.2 Special Session on GADRI

A special session on GADRI Achievements and Challenges was chaired by Prof. Hirokazu Tatano, Secretary-General, GADRI and Prof. Andrew Collins, Chair, GADRI Board of Directors. Four examples of activities were presented by GADRI member institutions.

Prof. Hirokazu Tatano and Prof. Andrew Collins

A Brief Introduction to GADRI was delivered by Prof. Hirokazu Tatano. GADRI aims to connect with institutes worldwide working in disaster risk reduction and disaster management research and practices. GADRI envisions enhancing disaster risk reduction and disaster resilience through exchange of ideas, research knowledge and networks in close collaboration with the member institutions thereby contributing to the Sendai Framework of Action and creating resilient societies.

GADRI Linkage to the World Bosai Forum 2017 was delivered by Prof. Yuichi Ono, International Research Institutes of Disaster Science (IRIDeS), Tohoku University, Japan. The World Bosai Forum[1] will be held in Sendai, Japan, from 25 to 28 November 2017. The aim of the World Bosai Forum to be held in Sendai Japan in November 2017 is bilateral and transdisciplinary approaches to discuss DRR with multi-stakeholders beyond national boundaries. Bosai is a traditional Japanese term indicating a holistic approach to reduce human and economic losses for disasters which represents activities in all disaster phases, including prevention, recovery, response and mitigation. Participation at the Bosai Forum is encouraged from private sector and others beyond academic and government sectors. The sessions will be on solution-oriented discussions on DRR with concrete examples provided by multi-stakeholders; promote the implementation of the SFA; provide access to business opportunities in DRR; and explore Japanese experiences on DRR and observe recovery process of the Tohoku Region. The conference venue is in the same place as where the World Conference on DRR was held, Sendai International Centre and the dates are from 25 to 28 November 2017. In addition to the World Bosai Forum, there will be two other back to back meetings: 2nd National Conference on Disaster Risk Reduction, 26–27 November, and the DRR Solution Exhibition 26–28 November which will take place at the same venue.

GADRI Capacity Development Activities were delivered by Prof. Wei-Sen Li, Secretary-General, National Science and Technology Centre for Disaster Reduction (NCDR).

The annual training courses cover diverse topics and look at problems from a different angle to understand how the society is impacted by disasters and how best to conduct successful evacuations, and communication mechanisms to contact people during an emergency. Last year, typhoon Megi hit central Taiwan during the training course on "natural disaster risk modelling and application", and the students received hands-on experience in emergency operations. The 2017 training course is on "evidence-based post-disaster recovery". The course will cover understanding process, procedures and difficulties during the recovery phase; science and technology support to develop a safer and sustainable environment at community level; and a field visit to the affected communities recovering from the 2009 typhoon Morakot.

Publication of State of the Art in Natech Risk Management was presented by Prof. Ana Maria Cruz, Head, Disaster Risk Management, DPRI, Kyoto University. Based on the recent workshop, natural disasters have continue to increase affecting many livelihoods in the Asian continent, while technological disasters are increasing

[1] World Bosai Forum 2017. https://www.worldbosaiforum.com/2017/.

too. This is due to the increase of chemical industries in Asian countries in the last 25 years. Natural and technological disasters are on the rise globally. Disasters triggered by technological hazards are mostly concerned with chemical accidents. These events e have impacts on community, economic development and environment. Natech disasters are high impact and low probability events. They generally fall outside of the proper safety management systems that are available and that are used by the chemical industry to evaluate the risks. The challenges are to address the interdependencies and possibilities of cascading events, such as for example as evidenced by the Fukushima Nuclear accident; hurricane Katrina; Awanchuan earthquake, and there are many other examples. Through GADRI members, we can benchmark methods and models through the different sectors and areas, and develop international risk management standards. Natech is relatively new and capacity building efforts are needed in all countries.

"**Sustainable Wadi Basins Development in Arid Regions: Innovative Technologies for Flash Floods Forecasting, Mitigating and Water Harvesting**" was presented by Dr. Sameh Kantoush, Associate Professor, Socio and Eco Environment Risk Management, DPRI, Kyoto University.

He introduced one of the ongoing GADRI transdisciplinary and region-specific projects. As the world population approaches 10 billion people, drought and water shortages increase. Water becomes one of the most critical and contested resources. Almost 80 countries have severe droughts, and the allocated water for their populations is very limited. In the Arab Region, between 2000–2010 drought frequency, one quarter of the total population were subjected to drought conditions for little less than two years within the 10 years, and inundation. During the last ten years, frequency and magnitude of flash floods have increased. Integrated flash flood management is needed for these areas. Through established networks in many different arid countries, the project on Flash Floods in the Wadi System continues reinforce its efforts to reduce disaster risk reduction by formulating research groups and capacity development activities.

1.3 Group Discussion Sessions

The group discussion focus was on four Priority Areas of the Sendai Framework for Disaster Risk Reduction 2015–2030. Participants had adequate time to interact and debate on their research and future activities. In order to facilitate the group discussions, a pre-survey on "Evaluating Current and Future Research Status and Identifying Most Important Research Themes" was conducted among the members of GADRI and fifty-four institutes participated in the survey, and identified nearly 250 current projects and 73 future projects.

Group discussion session – Hydrology and meteorology related group

In addition, each session chair and co-chair persons prepared a session concept note for their respective groups. The participants of the conference were requested to select the group they wish to participate in advance. Each of the following groups consisted of 30–50 participants and they actively contributed to the discussions. The results of the debates were presented during the two-panel sessions held on days two and three.

Day 1: *Group Discussion Session I: Deepening the Understanding of Disaster Risks*:

Session I—a—i: Hydrometeorology Related.
Session I—a—ii: Meteorology and Wind Related.
Session I—b: Earthquake, Volcano and Compound Disasters Related.
Session I—c: Geohazard Related.
Session I—d: Social and Human Science Related.

Day 2: *Group Discussion Session II*:

Session II—a: Enhancing Governance to Manage Disaster Risks.
Session II—b: Disaster Risk Reduction for Resilience.
Session II—c: Effective Response to Disaster Recover/Build Back Better.

Further details of each group discussion and conclusions can be found under the group discussion section.

Group discussion session – Earthquake and Volcano related group

Oral and poster presentation session provided opportunities, especially to young researchers and scientists to share their work and planned projects with the participants and to receive their inputs to improve the projects or move them to next levels.

Poster Session

1.4 Outcomes and Conclusion

The Third Global Summit concluded successfully and the following list a few of the decisive conclusions of the conference:

- Participation by scientists, engineers, policy makers and students
- Evaluation of current research and identifying gaps in most needed research areas.
- Participants' commitments to the four priority areas of the Sendai Framework for Disaster Risk Reduction 2015–2030.
- Commitment to dissemination information through the GADRI Book Series as book manuscripts.
- Compilation of a compendium to list GADRI members and their activities in research and capacity development activities.

The summit is a biennial event and the next summit will take place in March 2019. All members were requested to actively engage in the implementation of the targets of the Sendai Framework Agenda 2030.

GADRI Secretariat thank all participants, its members, the members of the Board of Directors, Disaster Prevention Research Institute, DPRI, Kyoto University, the

Kyoto University, City of Uji, Kyoto Prefecture and all those who have contributed to make the Third Global Summit a successful, enjoyable and a fruitful event.

References

GADRI ACTIONS 2017 https://gadri.net/resources/pdf/GANewsletter_3_April2017.pdf
Global Alliance of Disaster Research Institutes. https://gadri.net/
Sendai Framework for Disaster Risk Reduction 2015–2030. https://www.unisdr.org/we/coordinate/sendai-framework
World Bosai Forum 2017. https://www.worldbosaiforum.com/2017/

Chapter 2
Evaluating Current Research Status and Identifying Most Important Future Research Themes

Concept Notes for the Group Discussion Sessions

Hirokazu Tatano, Andrew Collins, Wilma James, Sameh Kantoush, Wei-Sen Li, Hirohiko Ishikawa, Tetsuya Sumi, Kaoru Takara, Srikantha Herath, Khalid Mosalam, James Mori, Fumihiko Imamura, Ryokei Yoshimura, Kelvin Berryman, Masahiro Chigira, Yuki Matsushi, Lori Peek, Subhajyoti Samaddar, Masamitsu Onishi, Tom De Groeve, Yuichi Ono, Charles Scawthorn, Stefan Hochrainer-Stigler, Muneta Yokomatsu, Koji Suzuki, Irasema Alcántara Ayala, Norio Maki, and Michinori Hatayama

Abstract This chapter focuses on group discussion sessions targeting the Priority Areas of the Sendai Framework for Disaster Risk Reduction 2015–2030. Day one group discussion session efforts were on Priority Area One—Understanding Disaster Risks; and Day two emphasis was on Priority Areas 2, 3 and 4.

Keywords Disaster risk reduction · Research gaps · Research · Disaster resilience · Build Back Better · Disaster preparedness

2.1 Introduction

The primary purpose being to share information on current and future research activities by the member institutes of GADRI and to share common agendas to align academic activities related to the four Priority Areas of action of the Sendai Framework for Disaster Risk Reduction 2015–2030 (SFDRR):

Priority 1: Understanding disaster risk.

H. Tatano · A. Collins · W. James (✉) · S. Kantoush · W.-S. Li · H. Ishikawa · T. Sumi · K. Takara · S. Herath · K. Mosalam · J. Mori · F. Imamura · R. Yoshimura · K. Berryman · M. Chigira · Y. Matsushi · L. Peek · S. Samaddar · M. Onishi · T. De Groeve · Y. Ono · C. Scawthorn · S. Hochrainer-Stigler · M. Yokomatsu · K. Suzuki · I. A. Ayala · N. Maki · M. Hatayama
Global Alliance of Disaster Research Institutes (GADRI), Kyoto University, Kyoto, Japan
e-mail: james.wilmatheonesta.6r@kyoto-u.ac.jp

Disaster Prevention Research Institutes (DPRI), Kyoto University, Kyoto, Japan

H. Tatano and A. Collins (eds.), *Proceedings of the 3rd Global Summit of Research Institutes for Disaster Risk Reduction*, Disaster and Risk Research: GADRI Book Series,
https://doi.org/10.1007/978-981-15-8662-0_2

Priority 2: Strengthening disaster risk governance to manage disaster risk.
Priority 3: Investing in disaster risk reduction for resilience.
Priority 4: Enhancing disaster preparedness for effective response and to "Build Back Better" in recovery, rehabilitation and reconstruction.

The aim of these discussion sessions was to explore disaster risk as a multi-dimensional concept; to share views on the most promising existing research; to find out about ongoing research gaps; and to find ways as how best to leverage existing knowledge on disaster risk to more effectively respond to disasters.

Priority Area 1—"Understanding' is related to pursuing deeper knowledge of disaster risks from various disciplinary perspectives and methodological approaches. Discussion Session I was arranged to facilitate debate within selected academic disciplines; sessions were categorized according to academic research groups.

In contrast to Discussion Session I, the Discussion Session II was arranged to focus on important research themes across academic disciplines along the Priority Areas of 2–4 identified in the SFDRR.

Preliminary results of the survey conducted among members of GADRI complemented discussion sessions. The survey covered the following points:

- How is current research contributing to disaster risk reduction?
- How are current research initiatives contributing to bridge the gaps between sciences, practice, and policy making?
- What are the missing areas of research that could fill the gaps?
- What are the most important research areas to be focused upon in the next 10–15 years?

Each of the group chairs received a table summarizing the current and future research activities and existing research gaps. The tabulation was divided among the four Priority Areas of the SFDRR, and identified academic disciplines in relation to the Priority Areas.

The discussion sessions helped to:

- To build a knowledge hub of disaster risk reduction and resilience to disasters including understanding of disasters, resilience to disasters, governance to manage risk, and response to Build Back Better
- To periodically assess and monitor progress for various initiatives
- To promote smooth access to evidence-based research results and data
- To identify effective mechanisms to collaborate with each other

2.2 Group Discussion Session I: Deepening the Understanding of Disaster Risks

2.2.1 Hydrometeorology Related

Sameh Kantoush, Wei-Sen Li, Hirohiko Ishikawa, Tetsuya Sumi, Kaoru Takara and Srikantha Herath

It is convinced worldwide that hydrometeorological extremes are the main driving force for most of devastating majority of natural hazards and this understanding provides a substantial rationale for deepening their understanding through detailed studies. Rapid increase of population, urbanization and economic developments haS pushed people for development in high disaster risk zones as floodplains and perhaps beyond will further increase hydrometeorology related risks. Understanding the hydrological characteristics such as spatiotemporal variability of rainfall, infiltration process, runoff generation, and surface and subsurface water interaction along with investigation of river basin geomorphology, topography, and geology as well as the impacts of climate change on increasing extreme event frequency and magnitudes, are the key components to reduce risks. This will be the optimal way to overcome such struggles. Participants were divided into small groups and discussed social and human aspects of disaster risks.

Topics discussed:

- Characteristics of extreme hydrometeorological events
- Reducing uncertainty of hydrological models forecasting and disaster risk reduction
- Understanding climate change impacts and reducing uncertainty
- Hydrochronological integrated approach for paleo, recent, and future floods
- Innovative approaches for understanding weather, river and hydrological processes and prediction
- Post-flood observation for the fragile system
- Monitoring, mapping, and indication of river basin support to take low-regret measures for protection against floods
- Cost-efficient solution to identify and protect hot spot areas
- Innovative radar and satellite techniques for rainfall monitoring and flood forecasting
- Coupling water and sediment monitoring and management in the river and flood plains
- Databases and networking.

S. Kantoush · W.-S. Li · H. Ishikawa · T. Sumi · K. Takara · S. Herath
Global Alliance of Disaster Research Institutes (GADRI), Kyoto University, Kyoto, Japan

Disaster Prevention Research Institutes (DPRI), Kyoto University, Kyoto, Japan

Key questions addressed:

- Why disaster impacts are increasing for atmospheric, and water resources disasters?
- How climate change affects hazardous processes?
- How science, technology and research address "new normal"?
- How scientific innovations are used in disaster risk reduction?
- How can science, technology, and research be applied to facilitate DRR collaboration between and among economies, the private sector, and international organizations?
- What are the research and knowledge gaps for atmospheric, hydrology, rivers, and water-related disasters?
- What will be needed in terms of technology, data, monitoring and assessment, modelling and analysis...etc.? (Interdisciplinary approaches)
- What are the specific study needs? (critical locations, frequencies, accuracies)
- What is the current availability, evaluation, and identification of knowledge gaps and data gaps?
- How to improve community understanding of the effects of flood hazards?
- How to upgrade the current design guidelines, protection measures, and implementation strategies?

Results and Recommendations
Estimating the edge of uncertainty:

- The uncertainty and confidence levels to approach decision makers and general public
- Balance between design and economics based on dialogue between researchers and decision makers (case from Vietnam flood in Ho Chi Minh City)
- Redundancy and coverage area: Partial system failure not full dis-function of the system
- Cascading factors of worse scenarios
- Fail-safe design
- Cost benefit analysis
- Investment on data science to reduce uncertainty of worse cases
- Public GIS Apps: to enhance risk understanding
- GADRI water forum, database cluster of communities to disseminate knowledge and exchange data, experiences related to hydrometeorology related
- Open up a communication media to exchange future activities and action plans on disaster reduction
- Case study "flood risk assessment project in Sri Lanka: Metro Colombo Flood Risk Assessment Project".

Strategies for fund raising

- The role of professional society (senior professional to form a group) and academy to influencing on the agendas of decision makers and big organizations (world bank)
- Professionals and societies direct the government priorities to think about new norm
- Private sector involvements for data set as well as financial resources
- Public awareness to increase investments
- Application program interface to manage different databases: Structural and non-structural data.

Strategies for data and data sciences

- Gaps between developing and developed countries and necessary collaboration to compensate each other
- Use of satellite data
- Exchange data, reliability and testing and related cost
- To set up new normal, we need evidence-based management and quality data
- International and transboundary rivers we have different languages (Standardization is needed)
- New normal: (evidence-based management to convince the decision makers why such normal is needed)
- New technology as unmanned aerial vehicle (UAV) as drone
- Compensate the data scarcity through satellite-based data.

Strategies for existing projects

- Identification of knowledge gaps and development of cross-discipline researches based on the existing 21 projects at DPRI
- Information sharing and give overview for such projects
- Good practice of implemented project for the counterpart countries
- Composing the matrix and define some indices for hydrometeorology to evaluate the existing projects
- Through the matrix, we identified gaps and suggested extensions for some of those projects.

Future directions

- Strategies for existing projects
- Identification of knowledge gaps and development of cross-discipline researches based on the existing 21 projects at DPRI
- Information sharing and give overview for such projects
- Good practices of implemented projects for counterpart countries
- Composing the matrix and define some indices for hydrometeorology to evaluate the existing projects.

Initiatives by GADRI

- GADRI water forum
- GADRI hydrodata archive/database system to share data for research on hydrological related disaster
- Create a cluster of communities to disseminate knowledge and experiences sharing related to enhancing disaster resilience
- Initiate a communication network to inform future activities and action plans on disaster risk reduction.

Proposals for mega scale development project

- Suggestions for international collaborative efforts
- Mega-scale development projects require special expertise and technical knowhow in multiple disaster risk reduction to overcome environmentally sensitive issues.
- A mechanism for facilitation of transferring the advanced knowledge and expertise in the form of direct training, forum discussions, sharing of training tools, and collective efforts established through collaborative activities.

2.2.2 Earthquake, Volcano, and Compound Disasters Related

Khalid Mosalam, James Mori, Fumihiko Imamura and Ryokei Yoshimura

The disaster risk reduction (DRR) aims at preventing and reducing socio-economic vulnerabilities to disasters, triggered by environmental and other hazards, e.g. earthquakes, tsunamis, floods, volcano-related, droughts, cyclones, etc. Therefore, DRR strengthens resilience of the urban communities and contributes to their sustainable development. It is now recognized that there is no such thing as a "natural disaster", but disasters following natural hazards.

Although the communities and nations around the globe are well aware of the risks due to the individual hazards (such as earthquakes or volcanoes) or multi-hazards, the socio-economic consequences of these hazards that would lead to disaster risk are rarely quantified. This is because the disaster risk depends not only upon the hazard(s), but also—from the built environment and its interactions with the social institutions of the urban communities—on economic and urban development, resilience of the buildings to multiple, cascading and interacting hazards, environmental sustainability, climate change, etc. Thus, the disaster risk is an indicator of the societal development and its management must involve multiple stakeholders,

K. Mosalam · J. Mori · F. Imamura · R. Yoshimura
Global Alliance of Disaster Research Institutes (GADRI), Kyoto University, Kyoto, Japan

Disaster Prevention Research Institutes (DPRI), Kyoto University, Kyoto, Japan

including social institutions, government, non-governmental organization, professional and private sector. To deepen the understanding of disaster risk, it is required to consider a multi-dimensional and people-centred approach. To this aim, several concepts were explored and discussed in this session:

Uncertainty Quantification

There are many sources of uncertainty in the processes that are used for determination of the effects of hazards on the built environment. As an example, for the earthquake hazard, these uncertainties exist in the definition of the hazard itself, in the response of a structure or a group of structures to the hazard, in the level of damage that can occur corresponding to the structural response and in the loss consequences of damage. Due to the inherent existence of these uncertainties, the most fundamental way of deepening the understanding of disaster risk is the development of robust methods for uncertainty quantification.

During the past few decades, there has been significant progress in understanding the occurrence and effects of damaging earthquakes. For example, there are long-term estimates of earthquake occurrence, detailed ground motion studies, scenario earthquake and tsunami models and probabilistic ground motion maps for many seismically active regions around the world. Moreover, the effects of the ground motions and tsunamis on the built environment have been studied by analyzing the damage from recent earthquakes, numerical modelling studies, and large-scale experiments.

Forecasting large volcanic eruptions has greatly progressed over the last 30 years through improvements of monitoring systems. These efforts have significantly reduced the loss of lives during recent large eruptions through timely warnings and evacuations.

As our understanding improves of the hazards and risks from earthquakes and volcanoes, there are needs for further efforts in implementing effective mitigation and education efforts. To this aim, some actions can be land use planning, informing people about risks, policy making and legislation, implementation of building codes.

Although such methods have been developed in the last two decades, e.g. the Performance-Based Earthquake Engineering (PBEE) methodology developed by the Pacific Earthquake Engineering Research (PEER) Centre, these methods are generally used by the academic community and not commonly used by professional engineers for solutions of real-life problems. Furthermore, such methods do not exist for other hazards such as volcanoes, extreme winds or for the case of multi-hazards or compound disasters. For the development and enhancement of such methods and for facilitating their wider spread usage, exchange of information and collaboration between GADRI member institutions is essential.

Build Back Better

In different parts of the world, consequences of hazards, e.g. earthquakes, are interpreted differently. In developing countries, the methods to avoid hazards from becoming disasters aim only at preventing the loss of lives, while the corresponding methods in developed countries aim at providing resiliency and post-earthquake functionality, which is one of the lessons from recent disasters such as 2011 Tohoku earthquake and tsunami. As the infrastructure in developing countries needs to be retrofitted to bring their performance to acceptable levels, in this upgrading process, the philosophy of only saving lives can be transformed to resiliency such that the planned infrastructure enhancements can have an updated resiliency objective. Such an objective transformation is only possible if sufficient economical resources are available. In this session, methods were sought to convince the decision makers of developing countries about the advantages of resiliency objectives in terms of environmental, social, and economic sustainable development, together with the necessity of corresponding policies. Furthermore, economical methods, such as low-cost base isolation systems, can be explored to make the resiliency objectives a realistic target in such countries. The use of stochastic financial methods to determine the feasibility of the use of low-cost protective systems and other methods was also discussed.

Community and City Scale Assessment
Deepening the understanding of disaster risk for a region or city is only possible considering the city or region as a whole instead of the sum of its components. Such city- or regional-scale disaster risk evaluation methods are very rare. In this session, we discussed collaboration opportunities and cross-institutional work that explore the development of regional scale disaster risk evaluation methods with robust uncertainty quantification incorporating holistically human systems, energy systems, environmental systems and urban systems.

Key topics discussed:

- Progress in seismic hazard maps and scenario earthquake and tsunami models
- Tsunami and earthquake early warning systems
- Improvement in seismic performance for buildings and infrastructure
- Intentions and changes of building codes and seismic regulations
- Improved prediction of volcanic eruptions and hazard maps
- Effects of volcanic ash on aviation
- Emergency response to earthquakes, tsunamis, and volcanoes
- Associated disasters (Landslides, fires, health and food security, and economic).

Key questions addressed:

- What are the research gaps that still need to be filled to improve our understanding of earthquake, tsunami and volcano hazards and risks?
- What should be the high priority efforts for current earthquake, tsunami and volcano disaster reduction efforts?
- What are the important groups that scientists and engineers need to communicate with for implementation strategies?
- What are the important tasks for scientists and engineers for public awareness and education efforts?
- What are the roles of scientists and engineers for improving laws and regulations related to these hazards and risks?

- What improvements are needed for response and recovery efforts following large disasters?

Results and Recommendations

1. **Research Gaps**
   - In Indonesia (and several other countries), there is a need for more basic information about locations of faults and magnitude of earthquakes.
   - Combination of earthquake and tsunami is important as a multi-hazard effect. If a structure is weakened by ground motion shaking, how will it be affected by a consequent tsunami?
   - There is generally a long time between large destructive tsunamis. Society changes during that time. Thus, there is a need to address changes of society during this long intervals between events (Group in PEER and USGS are currently working on Performance-based Tsunami Engineering).
   - Volcanic (even small phreatic) eruptions threaten tourists for popular volcanoes, e.g. in Indonesia or Japan.
   - There is a need for better monitoring of volcanoes, especially explosive ones.
   - Ash in atmosphere due to huge eruptions affect global climate and may lead to food crisis.
2. **High Priority Efforts**
   - Huge amount of modern data are becoming available.
   - Need to address uncertainties in the data.
   - Need to benefit from new methods and technologies, e.g. machine/deep learning or artificial intelligence, to quickly analyze data, even on a cell phone, to be able to make quick and informed decisions.
   - There is a need for improving the tsunami warnings for very large earthquakes. (Warning for 2011 Tohoku earthquake was underestimated).
   - There is a need to understand the distribution pattern of volcanic ash due to large eruptions, since this may result in large economic losses to the aviation industry.
3. **Important Groups for Communication**
   - Varies from country to country.
   - Need for better communication with the media. Currently, there may not be good communication, because of the differences in time scales and perspectives. Media wants quick information with some preconception. Scientists and engineers provide more long-term information along with the corresponding uncertainty.
   - Need for more technical skill capacity development (e.g. in numerical modelling or interpretation of results) in developing countries.
   - Education/training efforts are very important.

- Need for communicating with the construction industry (i.e. people performing the actual construction).
- Include information for general public.

4. **Important Tasks for Public Awareness and Education**

- In Haiti, social and political factors outweigh engineering problems. How to do engineering in such situations?
- In Morocco, there are many old and dangerous buildings in seismically hazardous areas. Where can funding come from to remedy this problem? Economic issues outweigh engineering problems.
- In Japan, how to convince people about extreme rare events?
- In Japan, scientists have largely given up on earthquake prediction; however, the public and policy-makers expect scientists to find prediction methods. There is no consensus on the expectations.
- The 2011 Tohoku earthquake was an important event and tsunami information was discussed during the session. First estimates of tsunami heights were underestimated in warnings, causing some people not to take action. However, some local people could evacuate before the warning following local traditional knowledge.

5. **Roles for Scientists and Engineers in Improving Regulation**

- There is a need to effectively communicate the meaning of risk.
- There is a need to show that *prevention* has economic advantage over *recovery*.
- Education efforts are important in all countries.
- There is a need to engage more with educators and policy-makers.
- People that scientist communicate with (government officials, policy-makers) often change very quickly (in a few years), so it is difficult to implement sustainable programs.
- In Africa, there are few earthquakes, so it is difficult to convey messages about earthquakes and volcanoes where there is little personal experience with such events. Students are a good starting point for education efforts.
- Building codes are often significantly advanced and adequate all around the world. Efforts are needed more on the enforcement aspect. In many countries, actual codes are quite good but not followed. Also, there exist problems related to corruption and compliance with building practices.

6. **Improvements Needed for Response and Recovery**

- Reminder: *Prevention is better than recovery economically and socially.*
- The World Bank helped recovery efforts in Haiti, Turkey, Pakistan.
- How can The World Bank and international organizations help improve hazard mitigation in developing countries?
- There is a need to communication with policy-makers at many levels.
  - What are we trying to achieve?
  - How to state the problem appropriately?

- Improve seismic safety of schools.
- Establish early warnings systems for storms (Tsunamis may be difficult).

2.2.3 Geohazards/Landslides and Surface Processes Related

Kelvin Berryman, Masahiro Chigira and Yuki Matsushi

This session addressed Priority Area 1 in the Sendai Framework for Disaster Risk Reduction that is deepening the understanding of disaster risk. This workshop was complementary to others in Session I and emphasized landslide and landscape change processes that impose hazards to downstream and downslope communities and infrastructure. In this session, we summarized current status of researches and identified future tasks in the fields of landslides, debris flow, and other types of geohazards (land surface erosion, liquefaction, settlement, rockfall, rockslide-induced dam/tsunami, etc.).

Geohazards exhibit various aspects with varying earth surface processes, depending on spatial distribution of topographic, hydrologic and geologic conditions. Geo-disaster potential is also time varying with the long-term phenomena such as regional tectonics and climate change, and by short-term processes, especially by heavy rainfall, snow melting, and earthquake shaking. A characteristic of this class of hazard is that the impact can sometimes be felt far from the initiating events, and thus, all of the temporal and spatial processes related to transport of debris also need to be understood. The session addressed general issues of defining maximum credible events, frequency–magnitude relationships in natural hazard perils, equivalent risk as a basis of prioritising risk reduction activities, and notions of acceptable risk.

Although spatial and temporal prediction of geohazards requires collaborative approaches across various scientific or engineering disciplines, there remain significant gaps between research outputs and present strategy for geo-disaster mitigation. We need interdisciplinary discussion for better understanding and modelling of earth surface processes related to geohazards with feedbacks by its application in social implementation.

Key topics discussed:

Inputs were sought from participants on the following topics:

- Characteristics of landslides—from rock avalanche to mud flow;
- Mechanisms of hill slope mass movements: current understanding and unsolved issues;
- Considering unique and compound drivers of landslide risk—earthquake, rainstorm, mountain instability, volcanic activity;

K. Berryman · M. Chigira · Y. Matsushi
Global Alliance of Disaster Research Institutes (GADRI), Kyoto University, Kyoto, Japan

Disaster Prevention Research Institutes (DPRI), Kyoto University, Kyoto, Japan

- Prediction for location, size, and timing of landslides and connection to debris flow mitigation;
- Case studies illustrating the types of impacts on communities and infrastructure;
- Assessing magnitude frequency distributions to quantify hazard and assess acceptable risk;
- Consideration of the effect of deglaciation on mountain region stability;
- Consideration of climate change modelling and how to model future landslide/landscape events so as to plan safely for urbanisation and infrastructure development;
- How to construct frequency–magnitude relationships for various perils and define maximum credible events;
- What constitutes acceptable risk;
- Geohazards in the changing world: strategy for the next decade.

Key questions addressed:

- What do we need to get better estimate of subsurface materials and structure?
- How can we identify and connect long-term and short-term factors?
- Do present-day technics change our views of geohazards? If yes, which one, how, and what will be the next steps?
- Can we evaluate real risk of geo-disaster in terms of hazards, exposure, and vulnerability?
- How can we achieve social implementation of geo-disaster mitigation?

Results and recommendations:

1. **How is current research contributing to disaster risk reduction?**
    - Progress is being made in understanding cascading geohazards
    - (e.g. Landslides to landslide dam formation to barrier lake breaching, or earthquake to debris sedimentation to scouring by subsequent rainfall);
    - Extensive use of new technologies–Lidar, InSAR, UAV's, geophysics, photogrammetry (e.g. structure for motion);
    - Landslide monitoring is advancing rapidly using these new techniques;
    - There are advances in interpretation and detection of pre-failure mechanisms, particularly rock avalanches and earthquake-induced failures;
    - Early warning for landslides based on displacements and rainfall monitoring is advancing;
    - Upscaling from individual landslides to catchment, and regional scale, using new technologies such as satellite observation (e.g. InSAR) is advancing;
    - Science input to policy is extremely variable. At present, this is more in post-disaster recovery but usually weak in risk reduction;
    - There are some good examples of evidence-based policy in some countries and in some sectors (e.g. building codes), but overall this is weak with a distinct lack of risk-based approaches to land use planning, infrastructure development, environmentally sustainable agriculture.

2. **How are current research initiatives contributing to bridge the gaps between sciences and policy making?**
    - Field interaction between researchers and local stakeholders is valuable;
    - Complex modelling of geohazard scenarios is now tractable with increasing computer capability—imaging in 3D, GIS, online data;
    - Working **with** stakeholders is improving along with 'translation' of research results **with** communities;
    - Increasing use of scenarios, risk rather than hazard assessment, and visualization and economic modelling are improving opportunities for inclusion of science in policy decisions;
    - Established laws and regulations have often been developed without a research basis and are therefore inhibiting rapid uptake of new data;
    - Sendai Framework has excellent **potential** for evidence-based policy development but there are many impediments.
3. **What are the missing areas of research to fill these gaps?**
    - Need to consider history of land use and landscape evolution in research projects;
    - Accounting for subsurface heterogeneity in landslide science
    - Coupling soil mechanics with hillslope hydrology and bio-activity to assess landslide potential;
    - Connecting long-term and short-term factors that leads to landsliding;
    - Temporal and spatial high-resolution monitoring of landslides;
    - There is a need for more scenario assessment and understanding of impacts;
    - Improved visualization as a means of communicating with stakeholders is needed;
    - Communication of **risk** and **life risk/economic impacts** is crucial to adoption of research results;
    - Globally, there remains a lack of consistent regional and local mapping at appropriate scales and of community-based mapping;
    - Hazard and risk education of all stakeholders is needed;
    - Early warning systems need to be reliable and appropriate for local needs;
    - Stakeholders must be involved in the establishment of early warning systems;
    - With exceptions (e.g. insurance, remote sensing), there is a lack of engagement by the private sector, but there are many opportunities especially for new technology and ICT partners;
    - ICT opportunities (e.g. cell broadcast can connect national level warnings directly with communities) but the 'audience' is not well understood and poor communication may cause additional stress;
    - Development of 'apps' enables optional engagement with information;
    - In general, scientists are not thinking enough about the 'audience' they are delivering to—there is a need for a tailored approach for **all** science communication;

- Communication of hazard and risk **must** include uncertainties/confidence measures;
- There is still work to do on effective use of social media.

4. **What are the most important research questions to be addressed in the next 10–15 years?**
   - Better understanding of the processes (e.g. seismic, rainfall), mechanisms, and thresholds of landslides: from rockfall to slides to flows;
   - Need to use probabilistic methods to acknowledge uncertainties—for example, for volumes and runout distance of landslides;
   - An improved ability to process and analyze 'big data';
   - In **every** nation, there is a need for a 'whole of society' discussion on acceptable levels of risk;
   - In many topic areas, there is a need to acknowledge that future states of the environment and hazards are different from the past—our data alone cannot forecast the future with confidence;
   - Policy development is a messy compromise between science, politics, economy, politics, and society—how to blend with transparency;
   - Need to re-assess laws and regulations to make sure they are enablers of DRR and not impediments;
   - Need urgent agreement between communities and regulators about whether to 'retreat or defend', especially on the coast.

2.2.4 Social and Human Science Related

Andrew Collins, Lori Peek, Subhajyoti Samaddar and Masamitsu Onishi

Hazards are considered to become disasters only when the people and communities are adversely impacted by them. Socio-economic characteristics of the people and communities, local political systems, group dynamics, cultural practices and land use patterns influence community disaster vulnerability and risk perception. Understanding social dimensions of disaster risk is therefore the cornerstone in deciding when, how and what preventive initiatives should be initiated. The concept of disaster risk has been variedly defined by different disciplines including for example sociology, psychology, anthropology, management and planning. It broadly covers the areas of risk perception, critical awareness, social vulnerability, household risk preparedness behaviours, social resiliency, livelihood risks, health risks and more. This points to the importance of identifying key areas of present research, potential research gaps or future research needs for disaster risk research from social and

A. Collins · L. Peek · S. Samaddar · M. Onishi
Global Alliance of Disaster Research Institutes (GADRI), Kyoto University, Kyoto, Japan

Disaster Prevention Research Institutes (DPRI), Kyoto University, Kyoto, Japan

human perspectives to better design and implement DRR strategies. The participants were divided into small groups to discuss the following social and human aspects of disaster risks.

Key topics discussed:

- Risk perception—social and cognitive aspects.
- Social vulnerability and resiliency—theories, indicators, and methodologies.
- Land use, city resiliency, and community planning.
- System networks and network dynamics.
- Early warning and evacuation.
- Health and livelihood risks in disaster.
- Social capital and the sense of community.

Key questions addressed:

- How are existing studies able to deliver knowledge in designing strategic actions for disaster risk reduction?
- What are the strengths and weaknesses of present studies in defining disaster risks for better disaster risk management?
- What are the roles of social scientists and planners in defining and managing disaster risks?
- What should be the future agenda and direction of study on disaster risk from social and human perspectives?

Results and recommendations:

1. **How current research is contributing to disaster risk reduction?**
    - There has been good work undertaken to start to plug what has been a major gap, particularly through 'human centred' research.
    - However, research emphasizing the social aspects of disaster risk reduction needs to be stronger to influence policy.
    - Focus has often been on building capacity.

1a. **Key points for consideration in progressing future research**

- There is a need to break the boundaries of boundary disciplines and turn DRR into 'new science'.
- There is a need to determine an appropriate framework to bring in various disciplines together, frame the DRR problem, then bring in the 'cavalry'. This can perhaps address the problem of fragmentation, disciplinary silos, duplication, etc.
- There is a need for feasibility studies to help mainstream research findings and recommendations into policy.
- There has been a lot done on the recovery phase, like BBB, but not much on planning

- How do you bring/apply DRR to real-life situations; link DRR to businesses? There is a need to get 'local' data, to better adapt 'solution' models to suit local needs and conditions.
- There is a need to understand local needs, for example, communication/mobile apps that best help people stay connected on disaster risk reduction needs.
- There is still a 'divide' where approaches/methods/practice are concerned, be it at various scales (national, regional etc.); perspectives (ontological domains) and concerns/goals/priorities.
- Prioritise research that is demand led.
- There is a need to provide and deeper understanding of the language of new forms of DRR.

2. **How current research initiatives are contributing to bridge the gaps between science and policy making?**

- There have been various scientific models and evaluation methods developed in the last decade.
- There has been a move to encourage grass root level applications, from DRR know how to application, BUT there is still a long way to go to developing a 'holistic' DRR.
- The gap between science and policy making is in part an issue of communication.

2a. **Key points for consideration**

- How do we measure 'contribution'? How do we know we have contributed? How are the research findings verified and who verifies them? Hazards are evolutionary in nature, and success is relative, e.g. time frame for recovery may not necessarily serve as a measure for success or effective contribution.
- There is a communication gap which has led to issues in:
  - Bridging studies of multiple scales;
  - Bridging perspectives/concerns/goals/priorities/know how/practices; and
  - Developing an interface between the scientific communities/think tanks with policy-makers, to help mainstream science.
- Avoid introducing complex techniques; hands-on practical approaches should be given focus.
- Break down government policy into practical approaches suited for civil society 'consumption'.
- There is a need to look at the relationship between research institutions and the government, e.g. in many developing countries, DRR may not be a priority and this will affect funding.

3. **What are the missing areas of research to fill in the gaps?**

- DRR and education—it is understudied. Need to review modules/curriculum.

- DRR speak—there is a need to translate and put into context DRR to various groups, experts to 'non-experts', to better communicate and make DRR speak effective and accessible. Take a close look at terms and terminology, understanding that cultures play a great influence on shaping understanding.
- Engagement—generational to professional.
- Wisdom and traditional knowledge which can add value if mainstreamed into research and policy making.
- Matching local data to help adapt solution models/approaches, tailoring it to suit a particular area/need/priority/capacity rather than a one-size-fits-all approach.
- Lack of understanding of how to underwrite risks.

4. **What are the most important research areas to be conducted in the next 10–15 years?**

   - Understanding local community livelihoods, in order to ensure their resilience.
   - Understanding the tangible and intangible aspect of disasters and DRR, particularly the terms associated which can be very subjective, taking into account the need to read in terms with human behaviour also, so that we can better underwrite the risks.
   - Linking risks and vulnerability with inequality.
   - Foster adaptive planning, to allow for innovation and flexibility and encourages inclusion of various stakeholder groups and understanding of communication issues.
   - Providing and understanding of how self-governance systems may function.
   - How best to respect values and follow ethical procedures in our socially oriented DRR research.

5. **Wrap-up points**

   - DRR research in theory can help tackle societal issues and problems.
   - Socially oriented research is crucial to enabling people centred approaches to DRR
   - There is a need to sort out the DRR house before it can be properly 'expanded' or 'operationalised'.
   - Some better established aspects of people centred DRR could broaden out to contribute to neighbouring areas.
   - Inclusion is key.
   - Upload papers for us to know each other and provide a platform for better communication among us.
   - Encourage easy access to GADRI network to catalogue knowledge, but members need to contribute papers.
   - Share the missing links and gaps to be filled by research, particularly areas to promote engagement/cooperation.
   - Mainstream wisdom

- Policies differ state to state; perhaps this can be catalogued and made available to promote understanding.
- Need to see whether our laws today are based on science.
- Time to reflect the voice of the local people and volunteers and give a platform to the 'voice of the survivors'.
- Time to consider self-governing systems that follow legal and ethical procedures in the use of technology.
- Always ask how our research matters to people.

2.3 Managing and Planning Disaster Risk Reduction

2.3.1 Enhancing Risk Governance to Manage Disaster Risks

Tom De Groeve, Masamitsu Onishi and Yuichi Ono

This session addresses priority 2 in the Sendai Framework, i.e. strengthening disaster risk governance to manage disaster risk. For the Sendai Framework, governance refers to defining roles and responsibilities, guide, encourage and incentivize the public and private sectors to take action and address disaster risk. According to the International Risk Governance Council (IRGC) white paper published in 2006, the concept of governance looks at risk-related decision-making when a range of actors is involved, requiring coordination and possibly reconciliation between profusion of roles, perspectives, goals, and activities. Of particular importance is the increasing role of the scientific community in disaster risk management. In reality, a variety of governance patterns DRR is observed in different contexts of hazards, geographical conditions and social norms. It implies the advantage of pursuing a governance pattern for DRR fitting for the hazard-specific and local-specific context rather than one-size-fits-all approach. To understand the variety in governance patterns, explicit knowledge on the compatibility between governance pattern and contexts is necessary.

Key questions discussed:

- What kinds of knowledge are necessary to understand the variety in governance patterns?
- How should we organize or coordinate activities of academic scientists to produce the necessary knowledge?
- How should hazard and local context be considered to strengthen governance?
- How to develop a better system to integrate scientific knowledge into the disaster risk reduction policy making process by each GADRI member country?

T. De Groeve · M. Onishi · Y. Ono
Global Alliance of Disaster Research Institutes (GADRI), Kyoto University, Kyoto, Japan

Disaster Prevention Research Institutes (DPRI), Kyoto University, Kyoto, Japan

Participants were allocated into sub-groups which consisted of about ten persons from different countries/regions and different research areas to identify important themes for DRR related to governance and possible platforms for promoting identified themes.

Results and recommendations:

Researches in disaster risk management (DRM) has contributed to;

- Development of national and local policy frameworks
- Establishment of agencies for disaster management
- Understand mechanisms of hazards –observation technologies, simulation technologies
- Technologies for making hazard map
- Real-time forecasting and early warning system
- Research in local engagement for implementation

Key points for consideration in future researches in DRM

- Research on science-policy interfaces in DRM
- Connecting science with decision-making
- Research in metrics for effectiveness of risk governance; evidence-based approach
- Sharing best practice and seeking affordable best practice
- Collaboration for collecting and sharing data
- Understanding of dynamic system of governance forms
- Adaptive management
- Need experts with different disciplines in hazards and academics backgrounds
- Management of relation to build trust and effective communication
- Scientific communicator, medias, education.

2.3.2 Disaster Risk Reduction for Resilience

Charles Scawthorn, Stefan Hochrainer-Stigler and Muneta Yokomatsu

There are currently a number of significant knowledge gaps and methodological issues that still impede effective disaster risk reduction (DRR) for resilience. A few examples suffice to demonstrate this point:

Key topics discussed:

A detailed comprehensive taxonomy of loss. Most DRR effectiveness is measured against the '3Ds' of "death, dollars and downtime", but many losses remain unaccounted for. HAZUS, the 2005 *Mitigation Saves* and other studies account for some

C. Scawthorn · S. Hochrainer-Stigler · M. Yokomatsu
Global Alliance of Disaster Research Institutes (GADRI), Kyoto University, Kyoto, Japan

Disaster Prevention Research Institutes (DPRI), Kyoto University, Kyoto, Japan

indirect losses, but many other types of losses are omitted. An exhaustive taxonomy of loss would provide a standard for loss estimation, identify important research gaps and assure vulnerable groups are adequately addressed.

GADRI Disaster Effects Portal (GDEP). The scientific method requires good data for model development and validation, which is currently lacking for disaster effects, the collection of which is ad hoc and inconsistent. GADRI supports development of a Disaster Effects Portal (GDEP) and is in an excellent position to foster disaster data collection on a globally consistent basis, via a network of disaster research institutes. However, no progress has been made on this front.

Measures of DRR effectiveness. The effectiveness of most DRR projects is inadequately assessed. Typical measures of effectiveness have all been borrowed from finance and include estimates of potential loss reduction irrespective of cost (simple gain), return on investment (ROI aka ROR), benefit-cost ratio (BCR) or internal rate of return (IRR). In all of these, the human dimension (when included) is typically translated into value of human life, and may or may not be temporally discounted. In this, the disaster field lags far behind the medical field, which has employed disability-adjusted life years (DALYs) for decades. Economic Adjusted Life Years (EALY) has been proposed to fill this gap, but has not gained traction. Research is needed into how to holistically measure DRR effectiveness.

Multi-facets of DRR and resilience. Global evidence repeatedly points out that risk is intricately connected to 'resilience' which is essentially a multi-faceted concept that is closely linked to the many aspirations of the sustainable development goals, including achieving food security, poverty alleviation, universal education and improved health and wellbeing. However, current DRR research focuses on resilience on only one or a few dimensions. There is the question how DRR can better be embedded within the multi-dimensional nature of resilience and overall development processes and goals. An interdisciplinary approach is certainly needed; however, how can this be actually achieved.

Models of Evaluating Long-Term Impacts of Disaster and Disaster Risk Reduction Policies

There exists a divergence in views regarding the impact of natural disasters on macroeconomic growth. There seems to be some consensus that natural disasters have a negative impact on short-term economic growth. On the other hand, the literature on the long-run effects of natural disasters is scant and inconclusive; some describe the expansionary disaster effects caused by, for example, Schumpeterian "creative destruction" while others note that the effect of creative destruction is limited. A contrasting conclusion is also found, stating that natural disasters have a negative long-term impact. It is indicated that compared to advanced societies, developing societies suffer from a greater magnitude of disaster damage in the long term. Moreover, the long-term effect of disaster mitigation or disaster risk reduction investments is a question that only a few economic growth models dealt with. It is also important to figure out conditions for "Build Back Better" in formalized way. It seems meaningful to categorize models (statistical model or simulation model, categorization with respect to assumptions on market conditions, variables, etc.), and link each type of models to appropriate objective of analyses and policy discussion.

Key questions discussed:

- How concepts of "resilience" are defined from various viewpoints in many academic and practical areas? What are relationships among those existing definition?
- What can be (a set of) proxies that reflects "resilience" of society? How can we measure and stock them to compose databases?
- Which kinds of models have been proposed to evaluate "disaster risk reduction for resilience"? Which kinds of approach are now required to be developed?
- What are the roles of infrastructure to enhance resilience in society?
- How do we facilitate cross-institutional work to fill in knowledge gaps and methodological issues that still impede effective disaster risk reduction (DRR) for resilience?

Results and recommendations:

The definition and concept of "resilience" varies depending on the problems faced:

- Different in hazard: flood, earthquake, landslide, as well as climate change
- Different in scales: communities, regions, and nations
- Different in society: developing ones or developed ones
- Different between engineering (functionality owing to smaller destruction) and social science (adaptability, flexibility, autonomous changes for survival)
- Risk (frequency)-based or consequence-based?

A commonly accepted, widely applicable terminology/taxonomy for resilience is needed and could be a cross-cutting project for GADRI. In this regard, some common factors and directions constraining "resilience" were identified:

- Related with critical infrastructure
- Must be defined with time span
- Must be linked with risk analysis
- Resiliency is a part of process, which changes in time when plans are updated by taking new factors into account.
- Social adaptability that depends on social governance structure matters.
- Local contexts matter to determine priority: development, needs, knowledge, etc.
- Should consider resetting of eco-system and community
- Resilience as a development dimension
- How to combine with other goals (Sendai Framework, sustainable development goals).

Other aspects of this discussion included:

- It is required to categorize three cases: (1) coming back to status quo, (2) reaching to something more (Build Back Better) or less, (3) repeating a stable cycle.

- It is demanded to define indicators that are operational and comparable, such as "resilience matrix". Indicators can be multi-dimensional while concept should be kept coherent to avoid overlapping.
- Ex-ante plan and ex-post plan should be different; the latter includes new factors such as changes of mind.
- Destruction and malfunction should be small in contexts of developing countries.
- Development that is associated with agglomeration may cause higher exposure of assets against disasters.
- It is important to better connect to local people (including school children) to find best solution.
- Society should be fast in action.
- A problem of implementation of mitigation investment is important to discuss.
- It is important to consider about the culture of community.
- It is a problem that improvement of building code in engineering is not linked with science.
- Especially in developing countries, transfer of knowledge is difficult. Moreover, life-cycle cost matters because some society cannot invest a lot in mitigation.

Based on this discussion, the following key needs were identified during the group discussion:

1. Technical needs

   - Need for modelling of recovery.
   - Introduce better hazard maps in developing countries.
   - Improve interdisciplinary community.
   - Develop system approach especially to cope with multiple hazards.
   - Formulate new matrices: more quantifiable treatment of resilience.
   - Develop new technologies to evaluate vulnerability.
   - Discuss on governance issues.
   - Learn traditional knowledge.
   - Classify properties of recovery of different stages.
   - Develop interoperability.
   - Understand dynamic nature of problems.
   - Involve local people in risk assessment.
   - Need for modelling the relations between resilience and development.
   - Need for broadening resilience to other development goals.

2. Risk communication-related needs

   - Risk communication should be developed to communicate vulnerability and capability of society.
   - Scenarios should be developed and shared to understand "what could happen?"
   - Researchers should communicate with policy-makers and local communities. For that purpose, "middle men" are necessary in between, such as NGO, in-house engineers.

- Politicians take an important role of institutionalization.
- Knowledge and participation
- Public should educate but also should learn from local experience.
- Use analogies, for example, illness.
- Show examples of past disasters what could happen.
- Public and local knowledge should be conveyed to researchers.
- Common language should be developed.
- Roles of local community leaders under limited resources should be identified.

2.3.3 Effective Response to Disaster Recovery/Build Back Better

Subhajyoti Samaddar, Koji Suzuki, Irasema Alcántara Ayala, Norio Maki and Michinori Hatayama

After the initial post-disaster response activities during the emergency and restoration periods, it is important to move quickly to the reconstruction and long-term recovery phase in order to restore a sense of normality in affected communities as soon as possible. Despite the increasing number of disaster experiences, post-disaster activities remain inefficient and poorly managed and need to be improved. Restoration of the damaged physical, social, economic, and environmental impacts of disasters is a complicated and drawn-out process. Reconstruction and recovery projects often focus on quick restoration of affected communities which can replicate and worsen existing vulnerabilities faced by the community. The built environment and infrastructure exactly as they were prior to a disaster often re-creates the same vulnerabilities that existed earlier. The reconstruction and recovery period following a disaster poses an opportunity to address and rectify vulnerability issues found in communities. Complete recovery requires attention to many different elements and effective coordination among the wide variety of stakeholders. Therefore, what the concept of Build Back Better (BBB) proposes is a broad holistic approach to post-disaster reconstruction in order to address the wide range of prevalent issues and ensure that the affected community is regenerated in a resilient manner for the future. To understand the fundamentals of BBB, it is valuable to analyze the complete list of propositions from all prominent studies and guidelines which depict effective post-disaster reconstruction and recovery. It is important to identify key areas of present research on and potential research gaps or future research needs for BBB.

The main objective of this session is to explore how the experiences on build back better by governments, communities, and organizations can strengthen the four elements of disaster recovery framework

---

S. Samaddar · K. Suzuki · I. Alcántara Ayala · N. Maki · M. Hatayama
Global Alliance of Disaster Research Institutes (GADRI), Kyoto University, Kyoto, Japan

Disaster Prevention Research Institutes (DPRI), Kyoto University, Kyoto, Japan

i. Policies and strategies,
ii. Institutional arrangements,
iii. Financing mechanisms, and
iv. Implementation arrangements and recovery management.

Key topics discussed:

The session discussion was guided by the following questions:

- How can past experiences better inform future implementation?
- How can reconstruction be linked to broader development goals and strategies?
- How can reconstruction be made more socially inclusive and non-discriminatory?

Key questions addressed:

- Risk reduction: reducing vulnerability and exposure
- Disaster recovery plan and resiliency
- Community planning and implementation science
- Decision support system for emergency preparedness
- Science and technology advisory boards
- Consensus building system and priority setting methodology among the stakeholders.

Results and recommendations:

1. "Build Back" is a critical step towards realizing disaster resiliency but the concept "Better" is value laden. Stakeholders vary in opinion about "Better" or successful Build Back options and policies. Because the risk perception and policy option are always contested.
2. There is no certain "risk" and uncontested or complete knowledge how to tackle the risks. Building Back "Better" is considered as relative, virtual, and comparative. Because of the value laden, subjective aspect of BBB, often stakeholders involve in conflict in the rehabilitation and reconstruction process.
3. To realize BBB, the critical question is to whose perspective of "Better" to be considered in the post-disaster reconstruction and rehabilitation process. It is the local community, especially the marginalized section, who remain unsolicited and isolated in the planning process.
4. For the successful implementation of BBB, following measures are suggested:

    i. A pro-active planning approach is necessary. The disaster risk management perspective should be incorporated from the designing stages of the reconstruction and rehabilitation. Often the disaster risk issues are considered in the end of the planning process or after disaster strikes.
    ii. The general objective Building Back Better is to achieve community's resiliency against future disaster. BBB therefore is an opportunity for local communities to rectify their past vulnerabilities and make a progress towards resiliency. Resiliency is conceptualized as bounced back to normal

life. But bounce back in the development process is a relative and contextual term. The standard of bounce back or resiliency that is applicable for Japan, maybe not applicable for Indonesia. What extent and how a community could bounce back depends on economic condition, social factors, and cultural beliefs of a particular community.

iii. Rehabilitation and reconstruction is a costly affair. Therefore, a community's capacity to bounce back should be based on local economy. Often the developing countries fail to bounce back because their BBB strategies do not focus on encouraging local economy but depend on foreign fund and aids. The regeneration of local economy is the key to ensure Build Back Better.

iv. Building Back Better is a long process. There should be different developmental goals and objectives in different phases of the BBB. We can divide the objectives BBB in three distinctive but sequential phases—NOW, SOON and sustainable.

v. In the developing country context, the BBB strategies should include reducing the livelihood risks. Without improving the livelihood security, Building Back Better will remain an illusion.

vi. The social-cultural factors play a critical role in disaster management. So the rehabilitation and reconstruction strategies should include social and cultural conditions of the local community. The model of building and planning strategies borrowed from the foreign countries did not work well because they do not satisfy the local social-cultural demands. Therefore, it is not uncommon that reconstructed buildings meet all the building standards of disaster resilient, but people refused to live in the reconstructed buildings as their cultural aspects are not included in designing the buildings.

vii. The local land use planning should be linked or integrated with the disaster risk reduction (DRR) strategies.

viii. The new development should be in-situate or relocation remains controversial. Some argued In situ development welcomes the prevailing risks in the area (e.g. settlement close to flood prone low-lying areas). Whereas some argued that people often refuse to move into relocated places because of livelihood insecurities, cultural and personal bonding and attachments with the area and so on.

ix. The cutting-edge technologies should be used and exploited to ensure steady and quick disaster recovery. But on the other hand, the technologies to be used in BBB should be pro-poor or affordable and energy efficient.

x. Building Back Better should be an integrated approach including different aspects of planning.

xi. Social networks of the communities should be used as a social infrastructure in the BBB process.

xii. There exist no groups or platform that advocates the integration of DRR strategies in the local planning process. For a successful implementation of BBB, it is important to have a group or platform like GADRI who can influence polices and planning strategies of the government. The advisory board should be formed at local, regional, and national level to promote integrated disaster risk management.

Part II
Selected Papers from Keynote Speeches

Chapter 3
Regional Science Engineering, Technology and Innovation (SETI) Support for Disaster Risk Reduction

Shahbaz Khan

Abstract Two important agreements, the 2030 Agenda for Sustainable Development Goals and Sendai Framework for Disaster Risk Reduction, are setting the stage for sustainable development leaving no one behind and actively pursuing strategies for disaster risk reduction at all levels. UNESCO Regional Science Bureau is working on pre-disaster actions especially by strengthening early warning systems, better building codes, capacity building and multi-hazard approaches for national policies. In this process, UNESCO is actively collaborating with other UN agencies and scientific entities to help underpin disaster risk reduction (DRR) with credible scientific data and analysis. At the regional level in Asia and the Pacific Region, UNESCO has developed a support strategy for international, interdisciplinary cooperation at community, national and regional levels. UNESCO actions bring together transdisciplinary actions between natural and social sciences, education, culture and communication areas. At the community level, the focus is on capacity building of resilient communities by utilising local and indigenous knowledge with smart communication technologies. The DRR capacities of UNESCO designated and affiliated sites (such as the World Heritage Sites, educational facilities, biosphere reserves and Global Geoparks) are being strengthened by contextualising DRR for sound integrated site management by fully complying with the principles of related programmes. In close cooperation with UNESCO's category-2 centres and chairs, the data gathering, scientific analysis and dissemination capacity in early warning systems are being improved in key affected countries such as Indonesia, Philippines and Pakistan. The Regional Science Bureau will continue to work in collaboration with its regional and global partners on the use of science and technology to support the member states' needs in disaster risk reduction and building sustainable resilience. This paper will focus on describing the regional support strategies in Asia and the Pacific Region and related key activities.

Keywords SETI · DRR · UNESCO · Asia and the Pacific · Partnerships

S. Khan (✉)
UNESCO Regional Science Bureau for Asia and the Pacific, Jakarta, Indonesia
e-mail: s.khan@unesco.org

H. Tatano and A. Collins (eds.), *Proceedings of the 3rd Global Summit of Research Institutes for Disaster Risk Reduction*, Disaster and Risk Research: GADRI Book Series,
https://doi.org/10.1007/978-981-15-8662-0_3

3.1 Introduction

Since 2015, global stakeholders including UNESCO are working on two important UN agreements the 2030 Agenda for Sustainable Development Goals and the Sendai Framework for Disaster Risk Reduction (SFDRR), 2015–2030. The SFDRR was signed in March 2015 by 187 countries in Sendai, Japan; the key aim of SFDRR is to reduce disaster-related losses of lives, livelihoods and health. The 2030 Agenda was signed in September 2015 outlining 17 goals with 169 targets covering a broad range of sustainable development issues such as ending poverty and hunger, improving health and education, making cities more sustainable, combating climate change and protecting oceans and forests while leaving no one behind.

Since the adoption of the Sendai Framework for Disaster Risk Reduction, 2015–2030, the UN system and related organisations have shown commitment to the implementation of the framework at all levels including formulation of regional support strategies such as by the UNESCO Regional Science Bureau for Asia and the Pacific (RSBASP 2017). The work of the UN General Assembly established an open-ended intergovernmental expert-working group, comprised experts nominated by states and is supported by the United Nations Office for Disaster Risk Reduction (UNISDR) now called UNDRR. This has helped involve relevant stakeholders, for the development of a set of possible indicators and terminology to measure global progress in the implementation of the Sendai Framework. In January 2016, UNISDR organised the Science and Technology Conference with a key outcome of the conference which was an updated Science and Technology Road Map 2030 to support the implementation of the Sendai Framework. A key further development in April 2016 was the UN Chief's Executive Board for Coordination endorsement of the revision of the UN Plan of Action on Disaster Risk Reduction for Resilience: Towards a Risk-informed and Integrated Approach to Sustainable Development.

UNESCO is an active collaborator of UNDRR (previously UNISDR) as part of the Global Alliance on Disaster Risk Reduction and Resilience in the Education Sector and the Resilient Cities Campaign at global and regional levels. A key aspect of the work is on bridging gaps science and society through promotion of Science, Engineering, Technology and Innovation (SETI) at the regional and global levels. The activities are shaped by the interface between the natural and social sciences, education, culture and communication, playing a vital role in constructing a global culture of resilient communities in a multidisciplinary manner.

The Regional Science Support Strategy for DRR for Asia and the Pacific Region focusses on building community resilience to natural hazards through early warning, capacity building, knowledge sharing and networking, and policy recommendations with the help and support for governments/local governments, civil society, research institutions, other UN agencies and international organisations, and all key stakeholders.

3.2 DRR and UNESCO Medium-Term Strategy, 2014–21

The UNESCO medium-term strategy (UNESCO 37C/4, 2014–21) maintained a major focus on disaster risk reduction (DRR) for most cost-effective means to mitigate the effects of disasters and save lives, heritage and infrastructure.

Emerging areas of DRR competence for UNESCO include:

- Remote sensing of emergency groundwater resources in drought-affected countries, national floods forecasting system and integrated flood management at the river basin level
- DRR education including through radio and other media
- Global expansion of tsunami early warning systems.

3.3 Regional Support Strategy on Science, Engineering, Technology and Innovation for Disaster Risk Reduction for Asia and the Pacific Region

Guided by the UNESCO 37 C/4), the Regional Science Bureau for Asia and the Pacific based at Jakarta developed regional support strategy on Science, Engineering, Technology and Innovation for Disaster Risk Reduction for Asia and the Pacific Region. The overarching mission of this strategy is "to advance the use of Science, Engineering, Technology and Innovation (SETI) to mitigate disaster risks and strengthen the resilience of societies through better understanding the hazard and risk, prevention and risk reduction, preparedness and early warning". As part of this strategy, mapping of SDGs and DRR was attempted, as shown in Fig. 3.1. This provides contextualisation of overall regional SETI programmes with DRR.

The key elements of the regional SETI DRR strategy are:

- Understanding disaster risk through evidence-based modelling and decision-making based on scientific data and local and indigenous knowledge as well as developing a database of DRR expertise in the region
- Strengthening DRR governance through institutional support
- Economic considerations for investments in DRR resilience
- Enhancing disaster preparedness for effective response, and to "build back better" in recovery, rehabilitation and reconstruction, this draws upon cooperation with architectural and engineering bodies in the region.

This strategy has guided several DRR activities integrating different disciplines; e.g. the natural sciences, education and communication/information sectors collaborate on the issue of safe schools as articulated in the Comprehensive Safe School Framework and UNISDR's Worldwide Initiative on Safe Schools. The Jakarta office in collaboration with UNESCO HQ has been working on provision of simple to use, science-based decision-making methodologies, like VISUS method (Visual Inspection for defining the Safety Upgrading Strategies) to ensure the structural safety of

Fig. 3.1 Mapping DRR in SDG targets in Asia and the Pacific Region

educational facilities for multi-hazards including earthquakes, floods, landslides and fires. This has also provided an opportunity for South–South cooperation as more than one hundred schools have been assessed in El Salvador and more assessments are currently being implemented in Laos, Indonesia, Peru and Haiti. Intergovernmental Oceanographic Commission (IOC) of UNESCO has established the Indian Ocean Tsunami Information Centre (IOTIC) in Jakarta supported by Government of Indonesia through the Agency for Meteorology, Climatology and Geophysics (BMKG).

3.4 Examples of Transdisciplinary DRR Activities

3.4.1 Strategic Strengthening of Flood Warning and Management Capacity in Pakistan

In response to the 2010 floods in Pakistan, UNESCO with financial support of the Government of Japan has developed a comprehensive strategy, to help the country develop and manage its responses to floods while upgrading the flood forecasting and early warning systems in the country. UNESCO offices in Jakarta and Islamabad have been working closely with Pakistan Meteorological Department (PMD), Pakistan Council of Research in Water Resources (PCRWR) and a number of other Pakistan partners and Japan organisations, UNESCO Category II Centre ICHARM (The International Centre for Water Hazard and Risk Management under the auspices of UNESCO) and Japan Aerospace Exploration Agency (JAXA) on the following aspects:

- Establishment of a technical foundation for sustainable capacity development on the flood management, forecasting, early warning and flood hazard analysis in Pakistan
- Technical studies to promote strengthening of cooperation with Indus river basin countries for transboundary flood management and transboundary data sharing
- Capacity building and education of communities on flood management for proper utilisation of flood hazard information and tools.

Further information on this project is available at: https://www.unesco.org/new/en/jakarta/natural-sciences/water-sciences/key-projects/pakistan-flood-project/.

3.5 Emergency Psychosocial Support for Secondary School-Aged Students Affected by Typhoon Yolanda in the Philippines

UNESCO Jakarta Office, in cooperation with the Philippines' Department of Education (DepED) developed a training manual and conducted a series of stakeholder trainings for the Project "Emergence Psychological Support for School-aged Students Affected by Typhoon Yolanda in the Philippines". The project yielded various lessons and best practices in implementing initiatives related to education in emergencies. The key learnings include active engagement of stakeholders in revision of teachers' manuals, wider engagement with local communities and ensuring post-project sustainability through mainstreaming in national curricula.

Project video can be accessed at https://www.youtube.com/watch?v=yILES1SJQ60.

3.5.1 Mobile Application for Preparedness

UNESCO in collaboration with donors (Government of Japan and USAID) and local stakeholders has developed Indus Water Manager, TANAH and SAI FAH as examples of educational gamification for disaster risk reduction in English and local languages. The cardboard-based and mobile apps provide integral lessons on, and reinforce the importance of, disaster preparedness, by exploring potential situations that may occur. Offered as platform-based games with various levels, users are provided with key survival lessons for all phases of disasters in an interactive manner.

3.6 Conclusions

The regional experiences show that DRR support and response strategies should be underpinned by sound SETI for ensuring access to freshwater, to education, to DRR information, to hazard assessments as well as to capacity building for multi-hazard disaster early warning systems and resource management at all levels. Multilevel stakeholder and policy engagement are critical for success of DRR actions. SETI support is critical for evidence-based multi-hazard DRR policy-making and timely implementation.

Acknowledgements The author acknowledges contributions of UNESCO staff especially Mr Ardito Kodijat, Dr Mee Young Choi, Dr Ai Sugiura, Dr Sachi Suzuki and Mr Raza Shah as well as partner organisations and their staff in shaping the Regional Science Support Strategy and Regional Support Strategy on Science, Engineering, Technology and Innovation for Disaster Risk Reduction for Asia and the Pacific Region.

References

UNESCO (2014) UNESCO's MEDIUM-TERM STRATEGY for 2014–2021 (document 37 C/4). https://www.unesco.org/new/en/bureau-of-strategic-planning/resources/medium-term-strategy-c4/

UNESCO Regional Science Bureau for Asia and the Pacific (2014) The Regional Bureau's science support strategy 2014–2021

UNESCO Regional Science Bureau for Asia and the Pacific (2017) UNESCO Regional Science Bureau support strategy for Asia and the Pacific: science, engineering, technology, and innovation for disaster risk reduction

Chapter 4
Science and Technology Commitment to the Implementation of the Sendai Framework for Disaster Risk Reduction 2015–2030

Virginia Murray

Abstract The Sendai Framework for Disaster Risk Reduction 2015–2030 was adopted by United Nations (UN) member states on 18 March 2015, at the World Conference on Disaster Risk Reduction held in Japan. The Sendai Framework went on to be endorsed by the UN General Assembly in June 2015. The Sendai Framework is wide in scope. This paper uses many resources of already published material to enable the reader to access a more complete summary of the science and technology commitment to the implementation of the Sendai Framework for Disaster Risk Reduction 2015–2030. In this paper on the role of science and technology engagement to provide evidence to inform policy and practice where possible, the author considered it important to emphasis the partnerships and learning she has been a part of and all significant statements that are included in this paper are in italicized quotes. The author is grateful for the many opportunities to engage at many levels with colleagues who also contributed so much to these opportunities for joint working and shared learning.

Keywords Sendai Framework · Disaster risk reduction · Scientific and technical · Science-policy · National level · Local level · Humanitarian emergencies

4.1 Introduction

The Sendai Framework for Disaster Risk Reduction 2015–2030 was adopted by United Nations (UN) member states on 18 March 2015, at the World Conference on Disaster Risk Reduction held in Japan. The Sendai Framework went on to be

V. Murray (✉)
Head of Global Disaster Risk Reduciton, Public Health England, England, UK
e-mail: Virginia.Murray@phe.gov.uk

UNISDR Scientific and Technical Advisroy Group (STAG) member 2008–2017, England, UK

Integrated Research on Disaster Rislk Reduction (IRDR) Scientific Committee, England, UK

H. Tatano and A. Collins (eds.), *Proceedings of the 3rd Global Summit of Research Institutes for Disaster Risk Reduction*, Disaster and Risk Research: GADRI Book Series,
https://doi.org/10.1007/978-981-15-8662-0_4

endorsed by the UN General Assembly in June 2015.[1] The Sendai Framework is wide in scope. It applies to the '*risk of small-scale and large-scale, frequent and infrequent, sudden and slow-onset disasters, caused by natural or man-made hazards as well as related environmental, technological and biological hazards and risks.*' (UNDRR 2015) The Sendai Framework promotes the shifting of the focus from managing disasters to managing risks. This requires a better understanding of risk '*in all its dimensions of vulnerability, exposure*' and hazards (UNDRR 2015). The Sendai Framework aims in principle to ensure that the multi-hazard management of disaster risk is factored into the implementation of the Framework at all levels as well as within and across all sectors.

The call has been made to enhance the scientific and technical work on disaster risk reduction in the Sendai Frameworks as follows:

- *To enhance the scientific and technical work on disaster risk reduction and its mobilization through the coordination of existing networks and scientific research institutions at all levels and in all regions, with the support of the United Nations Office for Disaster Risk Reduction Scientific and Technical Advisory Group, in order to strengthen the evidence base in support of the implementation of the present Framework; promote scientific research on disaster risk patterns, causes and effects; disseminate risk information with the best use of geospatial information technology; provide guidance on methodologies and standards for risk assessments, disaster risk modelling and the use of data; identify research and technology gaps and set recommendations for research priority areas in disaster risk reduction; promote and support the availability and application of science and technology to decision-making; contribute to the update of the publication entitled '2009 UNISDR Terminology on Disaster Risk Reduction'; use post-disaster reviews as opportunities to enhance learning and public policy; and disseminate studies;* (UNDRR 2017—Sendai Framework Paragraph 25 g).[2]

Stronger focus on science and building evidence (in its own right and to support scientific advice during preparedness, emergencies and recovery stages) invited engagement of the scientific community in the implementation of the Sendai Framework via the UNISDR Science and Technology Conference on the implementation of the Sendai Framework for Disaster Risk Reduction 2015–2030 held in January 2016, in Geneva Switzerland.[3] Over 600 organizations and institutions were listed by organization type, region and country as participants at the conference.[4] A UNISDR

[1]United Nations International Strategy for Disaster Reduction (UNISDR), Sendai Framework for Disaster Risk Reduction 2015–2030, available at https://www.unisdr.org/files/43291_sendaiframeworkfordrren.pdf. Accessed 02 August 2017.

[2]See Footnote 1.

[3]UNISDR Science and Technology Conference on the implementation of the Sendai Framework for Disaster Risk Reduction 2015–2030, January 2016. Available at http://www.unisdr.org/partners/academia-research/conference/2016/. Accessed on 31 July 2017.

[4]UNISDR Science and Technology Conference 27–29 January 2016, List of Organization Participants by organization type, region and country. Available at http://www.preventionweb.net/files/45270_listoforganizationsparticipatedinth.pdf. Accessed on 31 July 2017.

Science and Technology Road Map was undertaken to define the aspirations and concrete commitments of the science and technology community over the 2015–2030 period to support the delivery of the science needed under each of the four Sendai Framework priority areas of action which are:

- **Priority 1**: *Understanding disaster risk.*
- **Priority 2**: *Strengthening disaster risk governance to manage disaster risk.*
- **Priority 3**: *Investing in disaster risk reduction for resilience.*
- **Priority 4**: *Enhancing disaster preparedness for effective response and to 'Build Back Better' in recovery, rehabilitation and reconstruction'.* (UNDRR 2015 Sendai Framework Paragraph 20)

It was also identified that there was a need to develop ways to monitor progress and review needs.[5] Recognizing this, it was agreed that there was a need for peer review publications to report on the conference (Aitsi-Selmi et al. 2016; Dickinson et al. 2016). Following many detailed reflections on the way forward from the UNISDR Science and Technology Conference, Aitsi-Selmi et al. (2016) identified twelve key points with the following two points being of particular relevance to the Global Alliance of Disaster Research Institutes (GADRI) Second Global Summit of Research Institutes for Disaster Risk Reduction[6]:

- *'Need for formal 'national DRR science-policy councils/platforms' or a form of national focal points for science to support disaster risk reduction and management plans identified. Focal points could include platforms or chief scientific advisors function.*
- *'Conducting a periodic review of knowledge needs, new science (including implementation science), and research gaps. More effort is needed to work out how to achieve this and ensure avoiding duplication of effort'.* (Aitsi-Selmi et al. 2016)

In summary, there is now an opportunity for the scientific community and its institutions to meet the needs of an increasingly complex environmental and societal landscape to support the breadth of the aspirations to implement the Sendai Framework. This paper therefore reports on a selection of the actions taken by scientists at national, regional and global levels since the UNISDR Science and Technology Conference 2016 conference for the implementation of the Sendai Framework.

[5]UNISDR The Science and Technology Roadmap to Support the Implementation of the Sendai Framework for Disaster Risk Reduction 2015–2030. Available at http://www.preventionweb.net/files/45270_unisdrscienceandtechnologyroadmap.pdf. Accessed on 31 July 2017.

[6]Global Alliance of Disaster Research Institutes (GADRI) Second Global Summit of Research Institutes for Disaster Risk Reduction Available at http://gadri.net/events/2015/12/2015-2nd-global-summit-of-research-institutes-for-disaster-risk-reduction-development-of-a-research-.html. Accessed on 31 July 2017.

4.2 National Level Examples of Science and Technology Activities

Recognizing the identified call from the UNISDR Science and Technology Conference from the implementation of the Sendai Framework way forward of *'need for formal 'national DRR science-policy councils/platforms' or a form of national focal points for science to support disaster risk reduction and management plans identified'* a series of actions at national levels have developed. Examples at such work at national level include:

4.2.1 Christchurch, New Zealand

The People in Disasters conference was held in Christchurch on 24–26 February 2016, to coincide with the 5th anniversary of the major Christchurch earthquake of 2011, during which 185 people died and thousands were injured. As reported in the Concluding Statement of the conference '*Christchurch remains a city strongly affected by disaster with ongoing stressors many of them related to the prolonged recovery process*'.[7] Some of the key messages from the conference included the need for an all hazard, multidimensional approach for emergency planning for disaster risk reduction, response, recovery, rehabilitation and reconstruction strengthens the alignment with the global disaster risk reduction community and that mental health impacts were both immediate and latent—a longer term approach to mental health is needed.[8]

A Special Issue of the *Australasian Journal of Disaster and Trauma Studies* published a series of papers on: caring for companion animals; compassion fatigue among nurses; promoting Māori psychosocial recovery; family violence; managing diabetes post-disaster; comparing community recovery projects in Aotearoa/New Zealand and Japan; wellbeing of older people; and post-traumatic growth (Deely and Ardagh 2017). A report on the Christchurch earthquake mental health impacts and psychosocial recovery demonstrated that it was important to recognize the better understanding of the psychosocial health of the affected public, care providers and early responders, through the diverse voices heard (Hedlund 2016). In addition the Ministry of Health in New Zealand published a Framework for Psychosocial Support in Emergencies reflecting that distress is considerably more prevalent than mental health disorders and, for most people, the distress is tolerable, short-lived and depends on the duration of acute and secondary stressors and that psychosocial recovery is about positively adapting to a changed reality (Ministry of Health 2016). Finally, it

[7]Canterbury Clinical Network Transforming Health Care, Whanau Ora ki Waitaha. Concluding statement of the People In Disasters Conference, Christchurch, 2016 Available at http://ccn.health.nz/CCNMembersArea/tabid/1275/ArticleID/912/Concluding-statement-of-the-People-in-Disasters-Conference-Christchurch-2016.aspx. Accessed on 31 July 2017.

[8]See Footnote 7.

was determined that '*there was a need for another conference in 5 years to make sure the important work to recovery goes on and is properly documented for the Christchurch and the New Zealand communities but also for shared learning across the world*' (Aitsi-Selmi et al. 2016).

4.2.2 The Philippines

The 2016 Philippine National Health Research System (PNHRS) Week Celebration conference highlighted their growing commitment to disaster risk reduction (DRR). The event was led by the Philippine Council for Health Research and Development of the Department of Science and Technology (DoST) and the Department of Health and saw the participation of numerous research consortia from all over the Philippines. With a central focus on the Sendai Framework, the conference recognized the significant disaster risks faced in the Philippines while illustrating the strengths and experience in DRR. For example, key innovations in science and technology that were showcased include the web-based hazard mapping applications 'Project NOAH' (Lagmay 2012) and 'FaultFinder' (PHIVOLCS 2016), with another innovation being 'Surveillance in Post Extreme Emergencies and Disasters' (SPEED) (Department of Health 2011) which monitors potential outbreaks through a syndromic reporting system. Three areas noted for further development in DRR science and technology included: integrated national hazard assessment, strengthened collaboration and improved documentation. The combination of the risk profile of the Philippines, established national structures and experience in DRR, as well as scientific and technological innovation in this field are potential factors that could position the Philippines as a future global leader in DRR. The PNHRS conference saw the proposal to develop the Philippines into a global hub for DRR which was published in peer review paper entitled '*Developing the Philippines as a Global Hub for Disaster Risk Reduction—A Health Research Initiative as Presented at the 10th Philippine National Health Research System Week Celebration*' (Banwell et al. 2016).

4.2.3 UK

4.2.3.1 The UK Alliance for Disaster Research (UKADR)

In 2016, it was agreed that the UK Alliance for Disaster Research[9] would be developed. The primary motivation for this is to bring together the UK's rich and diverse

[9]UK Alliance for Disaster Research Available at http://www.ukadr.org/. Accessed on 1 August 2017.

disaster research community. The aim of the UKADR is that it '*will facilitate collaboration and partnership to aid representation of the research community at government level in the UK, and, where appropriate, help with the implementation of the Sendai Framework for Disaster Risk Reduction*'.[10] The Alliance is independent and managed by voluntary contributions from the UK research community and the current co-chairs are based at the University of Northumbria and King's College, London. The King's College Centre for Integrated Research on Risk and Resilience (CIRRR)[11] is an Integrated Research on Disaster Research (IRDR) Centre of Excellence[12] and the UKADR is part of their commitment to IRDR. The '*UKADR membership is open to staff and Ph.D. students of any research-and-capacity-building institute based in the UK that is active in researching disaster risk and its management. This includes universities and other educational institutions, charitable trusts, think tanks and research departments within government agencies, private businesses or civil society organizations*' (Banwell et al. 2016). For UKADR, it is agreed that disaster risk reduction, prevention and preparedness, disaster response, reconstruction and recovery are all areas of research interest that are applicable for membership. The first UK Alliance for Disaster Research Annual Conference was held on 9 January 2017 and 10 January 2017 at King's College London, with support from, Durham University and the Cabot Institute, University of Bristol, to help build the UK disaster science community and facilitate networking between all science traditions.

4.2.3.2 Natural Environment Research Council Science for Humanitarian Emergencies and Resilience (SHEAR)

Research funding is critical to support the acquisition of new knowledge. For example, the Natural Environment Research Council (NERC) is one of the UK's leading funders of independent research, training and innovation in environmental science. They invest public money in world-leading science, designed to help sustain and benefit from natural resources, predict and respond to natural hazards and understand environmental change. They work closely with policy-makers and industry to make sure that knowledge can support sustainable economic growth and wellbeing in the UK and around the world. They are supported mainly by the UK Department for Business, Energy and Industrial Strategy (BEIS), but their activities and funding decisions are independent of government. NERC works in partnership with other funders of research, for example, for the Science for Humanitarian Emergencies and Resilience (SHEAR),[13] a new international research programme jointly funded by

[10]See Footnote 9.

[11]King's College London Centre for Integrated Research on Risk and Resilience Available at https://www.kcl.ac.uk/sspp/research/cirrr/index.aspx. Accessed on 1 August 2017.

[12]Integrated Research on Disaster Research (IRDR) Centre of Excellence Available at http://www.irdrinternational.org/about/structure/icoes/. Accessed on 1 August 2017.

[13]Natural Environment Research Council. Science for Humanitarian Emergencies and Resilience (SHEAR) available at http://www.nerc.ac.uk/research/funded/programmes/shear/. Accessed on 31 July 2017.

the UK's Department for International Development, NERC and the Economic and Social Research Council. The overall SHEAR programme focuses on four areas: disaster risk assessment (mapping and analyses), sub-seasonal to seasonal forecasting, disaster risk monitoring and the integration of these into practical decision-making. The programme is targeting lower-to-middle income countries across sub-Saharan Africa and south Asia, focusing on the co-production of knowledge using a multi-disciplinary and problem-centred approach. The call for proposals closed in January 2016 and the four proposals have been funded until 2020 including two where I am on the advisory boards:

- Forecasts for AnTicipatory HUManitarian action (FATHUM).[14]
- Landslide Multi-Hazard Risk Assessment, Preparedness and Early Warning in South Asia: Integrating Meteorology, Landscape and Society (LANDSLIP).[15,16]

4.3 Regional-Level Examples of Science and Technology Activities

Regional-level science and technology has been inspiring in its commitments to developing implementation tools for the Sendai Framework for Disaster Risk Reduction. By using examples from the European Commission, the Pacific Community and from the Belt and Road Initiative led by China, it is possible to show case emerging developments.

4.3.1 European Commission

The European Commission has a range of important activities on disaster risk reduction and some of these are summarized below

[14]Research Councils UK FATHUM: Forecasts for AnTicipatory HUManitarian action 2016–2020 Lead Research Organisation: University of Oxford Available at http://gtr.rcuk.ac.uk/projects?ref=NE%2FP00041X%2F1. Accessed on 1 August 2017.

[15]Research Councils UK Landslide Multi-Hazard Risk Assessment, Preparedness and Early Warning in South Asia: Integrating Meteorology, Landscape and Society 2016–2020 Lead Research Organisation: Available at NERC British Geological Survey Available at http://gtr.rcuk.ac.uk/projects?ref=NE%2FP000649%2F1. Accessed on 31 July 2017.

[16]Research Councils UK Towards Forecast-based Preparedness Action (ForPAc): Probabilistic forecast information for defensible preparedness decision-making and action 2016–2020. Lead Research Organisation: King's College London Available at http://gtr.rcuk.ac.uk/projects?ref=NE%2FP000444%2F1. Accessed on 31 July 2017.

4.3.1.1 Disaster Risk Management Knowledge Centre, Joint Research Centre

The **Disaster Risk Management Knowledge Centre** (DRMKC) aims at enhancing the EU and Member States resilience to disasters and their capacity to prevent prepare and respond to emergencies through a strengthened interface between science and policy.[17] The DRMKC reflects the UN Member state call for the promotion of '*the use and expansion of thematic platforms of cooperation, such as global technology pools and global systems to share know-how, innovation and research and ensure access to technology and information on disaster risk reduction*'. (Sendai Framework paragraph 47c).[18]

The DRMKC, launched on 30 September 2015, is composed of several Commission services and a network of interested Member States and is part of the European Commission's Science Hub. It will be the focal point of reference to support the work of EU Member States as well as all actors involved in disaster risk management within and beyond the EU. The activities of the DRMKC are to support the translation of complex scientific data and analyses into usable information and provide science-based advice for disaster risk management policies, as well as timely and reliable scientific-based analyses for emergency preparedness and coordinated response activities. It brings together existing initiatives in which science and innovative practices contribute to the management of disaster risks.[19]

In the Executive summary of the Science for Disaster Risk Management 2017: Knowing better and losing less, led by DRMKC, it states that *reinforcing the science-policy interface should allow better exploiting and translating the complexities of scientific results into useful and usable policy outputs, through: efficient access and uptake of knowledge and research; a networked approach across relevant stakeholder communities; and continuous efforts towards innovation and new technologies and tools* (Poljanšek et al. 2017). The full report addresses the current status of disaster risk management and policy frameworks, the understanding disaster risk: risk assessment methodologies and examples; the need to understand specific disaster risk hazard-related risk issues including geophysical, hydrological, meteorological, climatological and biological risk as well as technological risk; with additional sections on communicating disaster risk, managing disaster risk and future challenges of disaster risk management (Poljanšek et al. 2017).

The report is an important achievement for the UNISDR Science and Technology Conference held on January 2016 call for '*Conducting a periodic review of knowledge needs, new science (including implementation science) and research gaps. More effort is needed to work out how to achieve this and ensure avoiding duplication*

[17]European Commission Disaster Risk Management Knowledge Centre (DRMKC) Available at https://ec.europa.eu/jrc/en/network-bureau/disaster-risk-management-knowledge-centre. Accessed 2 August 2017.

[18]See Footnote 1.

[19]Disaster Risk Management and Innovation Research Available at http://drmkc.jrc.ec.europa.eu/overview/About-the-DRMKC. Accessed 2 August 2017.

of effort.' This report is an important next step following on from the impact made in 2013 by the Intergovernmental Panel on Climate Change report on Managing the Risks of Extreme Events and Disasters to Advance Climate Change Adaptation made for encouraging the need for enhanced disaster risk reduction and management issues (IPCC 2012).

4.3.1.2 Examples of European Commission Research Programmes of Note for Disaster Risk Reduction

Community of Users on Secure, Safe and Resilient Societies

The European Commission has led on many research and capacity building programmes of major significance for the DRR community and developed initiatives to support their dissemination and implementation, e.g. through the Community of Users on secure, safe and resilient societies which is engaged in mapping EU policies and research projects funded by different programmes (under FP7 and H2020) for enhancing partnerships in H2020, and mapping the many programmes dealing with research, capacity building and education/training so a more a greater understanding of the programmes can be shared.[20] In summary, this group recognizes that '*in a world where the risks of man-made and natural disasters are ever-growing, the key question is how societies can enhance their resilience and become better prepared' (European Commission Community of Users).* The description of this work goes on to state that '*current threats, ranging from natural disasters to crime and terrorism, are posing challenges to the security of citizens, infrastructure and the environment*'; and '*In recent years, this topic has hence received increased attention from the European Union* (EU)' *(European Commission Community of Users).* The need to strengthen capacities in disaster risk/crisis management and improving resilience in the fields of CBRN-E (Chemical, Biological, Radiological, Nuclear and Explosive) and natural and man-made disaster management, as well as in the areas of border security and the fight against crime and terrorism, represent key EU policy and research challenges. Regular meetings are held with this Community and are very valuable to attend with much shared learning although the meetings are now live streamed with the outputs are published as summary documents.[21]

Specific examples of research programmes funded by the European Commission include.

[20]European Commission A Community of Users on Secure, Safe and Resilient Societies (CoU) Mapping EU policies and FP7 research for enhancing partnerships in H2020date NS. Available at https://www.securityresearch-cou.eu/sites/default/files/adaptivetheme/CoU%20Brief%20201705_05_The%20CoU%20Website.pdf. Accessed 2 August 2017.

[21]European Commission A Community of Users on Secure, Safe and Resilient Societies (CoU) Past Events Available at https://www.securityresearch-cou.eu/events. Accessed 2 August 2017.

PLAtform for Climate Adaptation and Risk ReDuction (PLACARD)

PLACARD's mission is to be the recognized platform for dialogue, knowledge exchange and collaboration between the Climate Change Adaptation (CCA) and Disaster Risk Reduction (DRR) communities and I am a member of their Advisory Board. The PLACARD interchange provides, in the large and complex landscape of stakeholder networks, research, policy initiatives and information sources, is working to enhance the coherence of and give direction to CCA and DRR research, policy and practices, strengthening cooperation and countering fragmentation between the domains. Thus, the programme is establishing a comprehensive coordination and knowledge exchange platform for multi-stakeholder dialogue and consultation to address gaps and fragmentation challenges, and support the development and implementation of an evidence base for research and innovation policies. The main work streams are the PLACARD interchange, stocktaking of the decision-making context, knowledge brokerage, institutional strengthening, evaluation, learning and legacy and communications & online platform. The programme partners are Fundação da Faculdade de Ciências da Universidade de Lisboa, Portugal, The Stockholm Environment Institute, Sweden and UK, Helmholtz-Zentrum Fuer Umweltforschung Gmbh, Germany, Centro Euro-Mediterraneo sui Cambiamenti Climatici Scarl, ITALY, UK Climate Impacts Programme, University of Oxford, United Kingdom, Umweltbundesamt Gmbh, AUSTRIA, University of Geneva, Switzerland, International Red Cross Red Crescent Centre on Climate Change and Disaster Preparedness and Alterra, Wageningen UR, The Netherlands. PLACARD is funded by the European Commission's Horizon 2020 research and innovation programme with the grant agreement No. 653255.[22] I am a member of the Advisory Board for PLACARD.

ANYWHERE (EnhANcing EmergencY Management and Response to Extreme WeatHER and Climate Events)

Extreme weather and climate events are the cause of a number of hazards affecting our society through their impacts on assets, and when interacting with exposed and vulnerable human and natural systems they can lead to disasters. According to the Global Assessment Report on Disaster Risk Reduction[23] Economic losses from disasters such as earthquakes, flooding, storm surges, wind storms, cyclones and tsunamis are now reaching an average of US$250 billion to US$300 billion each year, and two-thirds of them are due to extreme weather hazards such as flood, storm surges and windstorms. ANYWHERE is working to development of tools to support real-time coordination of the emergency response operations to face challenge of the

[22]PLAtform for Climate Adaptation and Risk reDuction (PLACARD) 2016–2020. Available at http://www.placard-network.eu/about-us/our-work/. Accessed 2 August 2017.

[23]UNISDR Global Assessment Report on Disaster Risk Reduction 2015 Making development sustainable: The future of disaster risk management Available at http://www.preventionweb.net/english/hyogo/gar/2015/en/home/index.html. Accessed 2 August 2017.

extreme weather and climate events. It employs cutting edge innovative technologies to develop systems to build a pan-European multi-hazard platform for faster analysis and anticipation of the risk prior the vent occurrence, improved coordination of the emergency actions and assist to raise the self-preparedness. The output is aimed at more effective early warning systems and decision support systems, accompanied by tailored online services developed to support self-preparedness, self-protection and self-response of citizens. The project has a wide range of project partners who are led from the coordination of Universitat Politècnica de Catalunya, Spain. I am a member of the Advisory Board. ANYWHERE is a EC-HORIZON2020-PR700099 programme.[24]

4.3.2 Pacific Community

The Pacific Community has shown major leadership in taking forward the science agenda for the implementation of Sendia Framework. For example, the Pacific Platform for Disaster Risk Management 2016 Outcome Statement identified health, science and technology particularly in the following paragraphs[25]:

*'**RECOMMEND** that Pacific Island countries and territories and partners take relevant steps to improve their understanding of climate and disaster risk, including through ….*

3.2 *Sound science and technology, including through capacity building and research, with the support of UNISDR Science and Technology Advisory Group (STAG) and explore the development of a Pacific Science and Technology Advisory Group to coordinate science and evidence base sharing for implementation.*
3.3 *Documentation and application of traditional and local knowledge systems*

*'**SUPPORT** the professionalization of the resilience sector through capacity building and the establishment of Pacific Regional Federation for Resilience Professionals (PRFRP).*

*'**ACKNOWLEDGE** the need for the region to follow International Health Regulations and support the WHO Safe Hospital Programmes'* (UNISDR Pacific Platform for Disaster Risk Management 2016 Outcome Statement)

By attending the Platform in Suva, Fiji and related meetings the need for a Pacific Science and Technology Advisory Group was clearly identified in the meetings around the Platform and the fact this was included in the Outcome Statement as a way ahead to regional disaster risk reduction and management health science and technology engagement across the region is invaluable.

---

[24]ANYWHERE (EnhANcing emergencY management and response to extreme WeatHER and climate Events) 2016–2020. Available at http://anywhere-h2020.eu/. Accessed 2 August 2017.

[25]UNISDR Pacific Platform for Disaster Risk Management 2016 Outcome Statement Available at http://www.unisdr.org/files/50790_ppdrm2016outcomestatement.pdf. Accessed 2 August 2017.

Documenting and applying traditional and local knowledge systems is also critical and when I was invited to the University of the South Pacific for a round table with the students, many reported on projects reflecting these topics.

Inclusion in the Outcome Statement of the professionalization of the resilience sector through capacity building and the establishment of Pacific Regional Federation for Resilience Professionals (PRFRP) is an invaluable achievement as well. The Pacific Community and the University of the South Pacific have worked on this topic for some time. By partnering with them and colleagues around the world, a supporting paper on accredited qualifications for capacity development in disaster risk reduction and climate change adaptation was published **(Hemstock et al.** 2016**)**.

In summary, it was identified that increasingly practitioners and policy-makers working across the globe are recognizing the importance of bringing together disaster risk reduction and climate change adaptation (CCA). It was noted that from studies across 15 Pacific island nations, a key barrier to improving national resilience to disaster risks and climate change impacts has been identified as a lack of capacity and expertise resulting from the absence of sustainable accredited and quality assured formal training programmes in the disaster risk reduction and climate change adaptation sectors. From the work undertaken at the UNISDR Science and Technology Conference in 2016 on the Implementation of the Sendai Framework for Disaster Risk Reduction 2015–2030, it was that much of the training material available are not reviewed either through a peer-to-peer mechanism or by the scientific community and are, thus, not following quality assurance standards. In response to these identified barriers, this paper focused on a call for accredited formal qualifications for capacity development identified in the 2015 United Nations landmark agreements in DRR and CCA and uses the Pacific Islands Region of where this is now being implemented with the launch of the Pacific Regional Federation of Resilience Professionals, for DRR and CCA.

A key issue is providing an accreditation and quality assurance mechanism that is shared across boundaries. By using the United Nations landmark agreements of 2015, support for a regionally accredited capacity development that ensures all countries can produce, access and effectively use scientific information for disaster risk reduction and climate change adaptation is needed and the newly launched Pacific Regional Federation of Resilience Professionals who work in disaster risk reduction and climate change adaptation may offer a model that can be used more widely.[26]

[26]Fiji News. Pacific Regional Federation for Resilience Professionals launched 17:05 Sun Oct 30, 2016 Available at http://trauma.massey.ac.nz/issues/2016-1/AJDTS_20_1_Hemstock.pdf. Accessed 2 August 2017.

4.3.3 *Belt and Road Initiative, China*

China's Belt and Road Initiative (BRI) is a call for an open and inclusive (mutually beneficial) model of cooperative economic, political and cultural exchange (globalization) that draws on the deep-seated meanings of the ancient Silk Roads (Liu and Dunford 2016). The BRI involves the establishment of a framework for open cooperation and new multilateral financial and scientific instruments designed to lay the infrastructural and industrial foundations to secure and solidify China's relations with countries along the Silk Roads.

In November 2016, I attended the First International Science Forum of National Scientific Organizations on the Belt and Road Initiative with 20 national scientific and research organizations from countries and regions along the Belt and Road Initiative. A declaration was made after a two-day forum attended by more than 350 scientists from 40 countries and regions.[27] The Beijing declaration, establishing an alliance of international scientists to promote cross-border cooperation highlights three areas of consensus

1. '*Strengthen cooperation in science, technology and innovation (STI) for the promotion of shared development*',
2. '*Build a platform of co-innovation and a long-term mechanism for STI cooperation*',
3. '*Focus on major challenges and organize related international programs*'.[28]

The Framework for the envisioned Belt and Road Initiative is illustrated in Fig. 4.1 and two examples of programmes that started in 2016 are summarized below.

4.3.3.1 Silk Road Disaster Risk Reduction (SiDRR)

As part of the scientific commitment to disaster risk reduction under the BRI, a 5-year international programme, Silk Road Disaster Risk Reduction (SiDRR), was launched to address issues related to hazards assessment and disaster risk reduction (DRR) was launched at a first workshop in November 2016. The workshop involved contributors from China and other countries including Pakistan, Nepal, Russian, Italy, UK, Sri Lanka and Tajikistan. Risk conditions on Belt and Road Countries (BRCs) were shared and science and technology advancements on DRR have been disseminated during the workshop. The programme agreed to adopt the objectives of Sendai Framework for Disaster Risk Reduction and Sustainable Development Goals and plans to assess disaster risk in BRCs and to propose measures for disaster risk

---

[27] Chinese Academy of Science. Scientists from Belt and Road Countries Gather in Beijing November 2016 Available at http://english.cas.cn/Special_Reports/CAS_in_the_Belt_and_Road_Initiative/News/201611/t20161118_170714.shtml. Accessed 2 August 2017.

[28] 'Beijing Declaration' Issued at the First International Science Forum of National Scientific Organizations on the Belt and Road Initiative (Nov. 7–8, 2016) Bulletin of the Chinese Academy of Sciences CAS Vol.30 No.4 2016. Available at http://english.cas.cn/bcas/2016_4/201703/P020170310587505348444.pdf. Accessed 3 August 2017.

Fig. 4.1 Framework of the envisioned Digital Belt and Road (DBAR) Initiative

reduction which can be appropriate both for specific sites in individual countries and more widely.

4.3.3.2 Digital Belt and Road (DBAR)

Belt and Road region represents the majority of the world's disaster deaths and losses. Given the rapid urbanization and continued population growth trend, the challenges faced by the region are only due to increase in the future. 'Digital Belt and Road' (DBAR) is an initiative that was started in 2016. As with the DRR initiative above the DBAR programme is in agreement with and in support of agreed global frameworks that include the Sendai Framework for Disaster Risk Reduction, the Sustainable Development Goals and the Paris Agreement on climate change. The aim is that under the DBAR framework, DBAR, jointly with Integrated Research on Disaster Risk (IRDR) International Programme Office (IPO), IRDR China National Committee (IRDR CHINA), Institute of Remote Sensing and Digital Earth (RADI), International Society for Digital Earth (ISDE) and CAS-TWAS Centre of Excellence on Space Technology for Disaster Mitigation (SDIM) conduct an international research programme. To do this, a DBAR Disaster Risk Reduction Working Group (DBARDISASTER) was formed to strengthen science capacities for sustainable development and disaster risk reduction, starting from year 2016. The First Consultative Workshop of the DBAR Regional Research Platform for Disaster Risk Reduction took place in December 1–2, 2016 in Sanya and a summary of this work is published as the Challenges of Disaster Risk Reduction in the Belt and Road: Contribution of DBAR (Fang et al. 2017).

Much ongoing work continues with the BRI for health, science and technology

4.4 Global Level Examples of Science and Technology Activities

Global-level science and technology developments have been inspiring in their commitments to developing implementation approaches for the Sendai Framework for Disaster Risk Reduction. By using examples from the many global organizations, institutions and research programmes such as the International Network for Government Science Advice (INGSA) and associated links including the InterAcademy Partnership (IAP), the International Association for Public Health Institutes (IANPHI), the WMO HIWeather programme, the COST Disaster Bioethics programme and Evidence Aid and their outputs, it is possible to see a few of the exciting developments in science, technology and health for the implementation of the Sendai Framework commitments that are now emerging.

4.4.1 The International Network for Government Science Advice (INGSA) and Its Links

The theory, practice and politics of scientific advice that build on the conclusions of an International Council of Science and its partners conference that was held in Auckland in 2014, led to the creation of the International Network for Government Science Advice (INGSA).[29] It is recognized that scientific advice to governments, as reported by INGSA, has never been in greater demand; nor has it been more contested. From climate change to cybersecurity, poverty to pandemics, food technologies to fracking, the questions being asked of scientists, engineers and other experts by policy-makers, the media and the wider public continue to multiply and increase in complexity. It is of note that at the same time, the authority and legitimacy of experts are under increasing scrutiny, particularly on controversial topics, such as climate change and genetically modified crops. It has been identified that the role of scientific advice and evidence features prominently in recent UN initiatives, such as the Sendai Framework on Disaster Risk Reduction, as identified in a peer review paper entitled '*Ensuring science is useful, usable and used in global disaster risk reduction and sustainable development: a view through the Sendai framework lens*' (Aitsi Selmi et al. 2016).

[29]International Network for Government Science Advice (INGSA). Available at http://www.ingsa.org/. Accessed 3 August 2017.

Indeed, partnership working for global agendas, through new collaborations like INGSA, the InterAcademy Partnership (Hassan et al. 2015) and the European Science Advice for Policy by European Academies (SAPEA) platform (Reillon 2016) are seen to be increasingly essential. Thus, in the international arena, there are now more regular and intense interactions between science advice, foreign policy and science diplomacy (Gluckman 2016). Several governments, including Japan, New Zealand, USA and the UK, have appointed science advisers to many of their domestic ministries and departments, but now increasingly it is seen as important that they also appoint science advisers to their foreign ministries.

In addition there has been debate about how to strengthen expert advice across the United Nations system, particularly in support of the sustainable development goals (SDGs), agreed by the UN General Assembly in 2015 (ICSU/ISSC 2015). As a result, a new UN Scientific Advisory Board was established in 2014, and a recent review calls for its remit to be expanded by the incoming UN Secretary General (UN SAB 2016).

INGSA's second international summit in Belgium in September 2016 was organized in partnership with the European Commission and addressed Science and Policy-Making: towards a new dialogue. This meeting brought together around 450 experts from seventy countries to debate the state of the art in scientific advice and evidence informed decision-making, across a variety of national and international policy. In the assessment of the *Cool heads in crises: How to provide timely advice in emergencies where* oil spills, epidemics, earthquakes, nuclear disasters, financial crises and food safety scares were discussed. it was noted that experts may be called upon to explain and respond to emergencies in real time (European Commission 2017). It was agreed that science advice is often viewed as a slow, deliberative process feeding into a complex political system, but, when disaster strikes, the 'rules of the game' change in an instant. It was considered that the timeframe for analysis and advice shortens dramatically and experts are sometimes thrust into the limelight to make public statements or predictions. These pressures can reveal and exacerbate stresses and deficits at the science–policy interface and require the development of robust risk management structures. Suggested solutions included consideration that crisis response: requires stronger science advice systems and to maintain trust when dealing with crises, national authorities should develop, as recommended by OECD in 2015:

- Permanent structures or mechanisms
- A central clearing house and contact point
- Clear reporting processes
- A predefined public communication strategy
- International coordination.[30]

[30]Scientific Advice for Policy Making, OECD (2105) Available at http://www.oecd-ilibrary.org/science-and-technology/scientific-advice-for-policy-making_5js33l-1jcpwb-en. Accessed 3 August 2017.

Ongoing meetings on evidence informed policy-making continue, including a recent one organized recently by Organisation for Economic Co-operation and Development (OECD) Directorate for Public Governance in cooperation with the European Commission's Joint Research Centre (JRC), the Campbell Collaboration and the International Network for Government Science Advice (INGSA) in June 2017.[31]

4.4.2 International Association of National Public Health Institutes

The International Association of National Public Health Institutes (IANPHI) links and strengthens the government agencies responsible for public health, an essential science and organizational resource in many countries for the delivery of the Bangkok Principles for the implementation of the health aspects of the Sendai Framework for Disaster Risk Reduction 2015–2030 that were recommended at the International Conference on the Implementation of the Health Aspect of the Sendai Framework for Disaster Risk Reduction 2015–2030, held on 10–11 March 2016.[32] In summary, IANPHI improves the world's health by leveraging the experience and expertise of its member institutes to build robust public health systems with 108 members from 93 countries (and growing), benefiting more than 5 billion people on 4 continents (International Association of National Public Health Institutes (IANPHI) 2017).

At the IANPHI's 2016 Annual Meeting in Shanghai, October 17-21 awareness was raised new of global policy and action towards disaster risk reduction and possible engagement opportunities for NPHI following a technical session titled *'Engaging NPHIs and IANPHI in Disaster Risk Reduction'*.[33] The session was moderated by Duncan Selbie, Chief Executive of Public Health England, and it focused on advocacy and implementation of the Sendai Framework for Disaster Risk Reduction 2015–2030. As discussed, the framework evolved from the need to ensure that disaster risk reduction policy reflects an understanding of the complexity of disaster risk in the twenty-first century and includes health as a key outcome. Implementation calls for a multi-sectoral, transdisciplinary and collaborative all-hazards approach to prevent, prepare for, respond to and recover from disasters. It provides an opportunity to align with twenty-first century global public health strategies through evidence-based policy and action—a key point of synergy with IANPHI and the mission of

[31] Organisation for Economic Co-operation and Development (OECD) Governing Better Through Evidence-Informed Policy Making June 2017 Available at http://www.oecd.org/gov/evidence-informed-policy-making.htm. Accessed 3 August 2017.

[32] UNISDR/World Health Organization. Bangkok Principles for the implementation of the health aspects of the Sendai Framework for Disaster Risk Reduction 2015–2030 available at: http://www.who.int/hac/events/2016/Bangkok_Principles.pdf. Accessed 3 August 2017.

[33] International Association of National Public Health Institutes (IANPHI). IANPHI Forms New Disaster Risk Management Group 2016. Available at http://www.ianphi.org/news/2016/disasterriskmanagement.html. Accessed 3 August 2017.

the member NPHIs (Aitsi-Selmi and Murray 2015; Aitsi-Selmi and Murray 2016). The session resulted in a call to encourage IANPHI and member NPHIs to:

- Advocate on the centrality of health to emergency and disaster risk management
- Incorporate the health aspects of the Sendai Framework within national emergency and disaster risk management policies and measures
- Strengthen collaboration with ministries of health and other stakeholders to address emergency and disaster risk management
- Strengthen the evidence base for health emergency and disaster risk management.

4.4.3 WMO HIWeather Programme

With the agreement of the Global Targets and the related indicators the science community is called to respond to the Sendai Framework for Disaster Risk Reduction's target g) to '*substantially increase the availability of and access to multi-hazard early warning systems*' in the context of weather-related hazards. The World Meteorological Organization has responded to this call with its partners to develop the High Impact Weather (HIWeather) project, a ten-year programme starting in 2016, with its mission to '*Promote cooperative international research to achieve a dramatic increase in resilience to high impact weather, worldwide, through improving forecasts for timescales of minutes to two weeks and enhancing their communication and utility in social, economic and environmental application'.*[34]

Recent developments in weather forecasting have transformed our ability to predict weather-related hazards, while mobile communication is radically changing the way that people receive information. At the same time, vulnerability to weather-related hazards is growing through urban expansion, population growth and climate change. HIWeather has identified and is promoting research in key multidisciplinary gaps in our knowledge, including in basic meteorology, risk prediction, communication and decision-making that affect our ability to provide effective warnings.

The results showcase leading edge capability aiming to build developing country capacity. In summary, a community that knows the hazards it faces is warned when a hazard approaches, and knows what to do when warned, and is a more resilient community. Thus, HIWeather aims to:

- Improve weather forecasts through advanced observing, data assimilation, modelling and ensembles
- Improve hazard forecasts by closer coupling of weather and hazard models
- Develop impact forecasts, based on each community's vulnerabilities
- Develop improved ways of communicating warning information
- Demonstrate the value of better warnings.

---

[34]World Meteorological Organization. High Impact Weather Project (HIWeather) 2016. Available at https://www.wmo.int/pages/prog/arep/wwrp/new/high_impact_weather_project.html. Accessed 3 August 2017.

4.4.4 *COST Disaster Bioethics Programme—2012–2016*

Disasters raise many ethical dilemmas. Decisions must be made about who receives the limited resources, sometimes with life-and-death consequences. Restrictions may be required to limit the spread of disease or maintain security, but may also interfere with people's rights. Conflict and violence add further complexities, as does taking care of displaced persons and refugees. Yet responders are often provided little or no training to help address such ethical dilemmas, which can lead to moral distress. To address these and other ethical issues, the Disaster Bioethics Action was funded from 2012 to 2016 by the COST IS 1201 Association and supported by the EU Framework Programme Horizon 2020. The network has 28 COST countries participating with over 120 members from academia, non-governmental organizations and international organizations.[35]

In January 2016, at the request of UNISDR and UNESCO, the Action helped to organize an ethics event in Geneva at the implementation conference for the Sendai Framework for Disaster Risk Reduction 2015–2030. The Geneva event identified the need for explicit, ongoing ethics reflection and guidance in many areas of disaster risk reduction. Thus, while the Action has contributed to developing the new field of disaster bioethics, much remains to be done to promote human rights and ethical action during disaster planning, responding and research,

Many publications from the COST Disaster Bioethics programme resulted from the project and of note these include a UNISDR case study (O'Mathúna and Von Schreeb 2015), contributions to the World Health Organization 2016 Guidance for managing ethical issues in infectious disease outbreaks (WHO 2016) and a recent article on research ethics and evidence for humanitarian health (O'Mathúna and Siriwardhana 2017).

4.4.5 *Evidence Aid*

Evidence Aid was established following the tsunami in the Indian Ocean in December 2004. It uses knowledge from systematic reviews to provide reliable, up-to-date evidence on interventions that might be considered in the context of natural disasters and other major healthcare emergencies. Evidence Aid seeks to highlight which interventions work, which does not work, which need more research, and which, no matter how well meaning, might be harmful; and to provide this information to agencies and people planning for, or responding to, disasters. The Mission Statement of Evidence Aid is 'To inspire and enable those guiding the humanitarian sector to apply an evidence-based approach in their activities and decisions'.[36] Evidence Aid links to the Cochrane Library (ISSN 1465-1858) which is a collection of six

---

[35]Disaster Bioethics COST Action IS1201 2012–2016: Disaster Bioethics: addressing ethical issues triggered by disasters. Available at http://disasterbioethics.eu/. Accessed 3 August 2017.

[36]Evidence Aid Available at http://www.evidenceaid.org/who-we-are/. Accessed 4 August.

databases that contain different types of high-quality, independent evidence to inform healthcare decision-making, and a seventh database that provides information about Cochrane groups.[37]

Evidence Aid resources are split into four categories: health issues, emergency type, humanitarian cluster and person groups. In addition to the resources, there are also five disaster-specific collections. Three are available through the Cochrane Library for the topics—burns; post-traumatic stress disorder; and flooding and poor water sanitation). Four special collections are available directly from the evidence aid website—Windstorms, Earthquakes, Ebola, The Health of Refugees and Asylum Seekers in Europe and Zika.[38]

4.5 National Targets and Indicators of the Sendai Framework

Adoption of the Sendai Framework by the UN Member States includes agreement on seven global targets to assess global progress in disaster risk reduction. The Sendai Framework states that: 'these targets will be measured at the global level and will be complemented by work to develop appropriate indicators. National targets and indicators will contribute to the achievement of the outcome and goal of the present Framework'. (Sendai Framework Paragraph 18).[39]

The need to support UN member states is also identified in their delivery of UN General Assembly Resolution A/71/644, 'Report of the open-ended intergovernmental expert working group on indicators and terminology relating to disaster risk reduction' (adopted on February 2 2017) (United Nations 2017). This resolution describes the set of indicators to measure global progress in the implementation of the Sendai Framework (Annex 1) and provides the agreed update of the publication entitled '2009 UNISDR Terminology on Disaster Risk Reduction'. Currently technical guidance for countries on indicators, methods and data required for is being prepared for the Global Targets.

The Sendai Framework addresses the need for a shared understanding of disaster risk through its global targets and indicators, which also provide a focus for attention on concerns that are universal to all countries (WHO (World Health Organization) 2015). These targets can act as a catalyst to accelerate change within countries as their high public profile attracts political commitment and financial resources. The benchmarking of targets as a global process can also provide a strong motivating factor for countries. Nevertheless, indicators are not without their limitations. Their measurement often relies on robust and complete data, which may not be available across all countries, resulting in comparability issues. Moreover, indicators have the

---

[37]Cochrane Library Trusted evidence, Informed decisions, Better health. Available at: http://www.cochranelibrary.com/about/about-the-cochrane-library.html. Accessed 4 August 2017.

[38]Evidence Aid Resources, available at http://www.evidenceaid.org/resources/. Accessed 4 August.

[39]See Footnote 1.

potential to be misleading, if the data, assumptions, or analyses behind them are incorrect. Aggregated data, for example, may mask inequalities within vulnerable groups that, unless disaggregated, will remain hidden to policy-makers (Maini et al. 2017).

A UNISDR/Integrated Research on Risk and Resilience/Public Health England workshop on Disaster Loss Data took place in February 2017 at the Royal Society in London, convened following the work completed by the OIEWG on Terminology and Indicators Relating to Disaster Risk Reduction and sought to further develop the loss data technical guidance notes concerning the Global Targets. The discussions at the workshop resulted in a number of proposals for changes and revisions within the technical guidance notes. The UNISDR Secretariat is to review all proposals and will ensure that all changes made to these notes are fully aligned with the discussions of the OIEWG, and in full compliance with General Assembly Resolution A/71/644 (Fakhruddin B, Murray V, Maini R. Disaster loss data in monitoring the implementation of the Sendai Framework ICSU and IRDR. Available at https://www.icsu.org/cms/ 2017).

Examples of other resources that contribute to supporting the work to deliver the indicators include:

- Integrated Research on Disaster Risk (IRDR)[40] and its Disaster Loss Data (DATA) project[41] which are part of a decade-long research programme co-sponsored by the International Council for Science (ICSU), the International Social Science Council (ISSC) and the United Nations Office for Disaster Risk Reduction (UNISDR). It is a global, multi-disciplinary approach to dealing with the challenges brought by natural disasters, mitigating their impacts and improving related policy-making mechanisms. Core funding for IRDR is provided by the China Association for Science and Technology. IRDR International Programme Office is hosted by Institute of Remote Sensing and Digital Earth (RADI) Chinese Academy of Sciences. By working with the International Council for Science: Committee on Data for Science and Technology (CODATA)[42] and its Linked Open Data for Global Disaster Risk Research[43] which aims to provide guidance that is of value to national policy for disaster data needs to follow the agreements by the UN Member States after they have determined how they will provide the agreed indicators for the global. A report entitled Gap Analysis on Open Data Interconnectivity for Disaster Risk Research is being completed.

[40]Integrated Research on Disaster Risk (IRDR). Available at http://www.irdrinternational.org/. Accessed on 4 August 2015.

[41]Integrated Research on Disaster Risk (IRDR) Disaster Loss Data (DATA) project. Available at http://www.irdrinternational.org/projects/data/. Accessed on 4 August 2015.

[42]International Council for Science: Committee on Data for Science and Technology (CODATA) http://www.codata.org/. Accessed on 4 August 2015.

[43]International Council for Science: Committee on Data for Science and Technology (CODATA) Linked Open Data for Global Disaster Risk Research. Available at http://www.codata.org/task-groups/linked-open-data-for-global-disaster-risk-research. Accessed on 4 August 2015.

- The UN Sustainable Development Solutions Network—Data for Sustainable Development—is part of the data revolution which is poised to transform the way governments, citizens and companies do business.[44] The revolution is being defined by the explosion in availability of data resources and rapidly evolving technologies, which are changing the way data is collected, processed and disseminated. A report entitled 'A World that Counts'[45] provided a link from data for development to data for disasters. The United Nations E/CN.3/2016/2 Economic and Social Council Report of the Inter-agency and Expert Group on Sustainable Development Goal Indicators has now brought together commonalities between the Sendai Framework and the SDG indicators[46] ongoing work to collaborate continues with many opportunities for continuing joint partnership working and activities to take forward the implementation of the global targets of the Sendai Framework

4.6 In Summary

Since the UNISDR Science and Technology Conference on the implementation of the Sendai Framework for Disaster Risk Reduction 2015–2030 held in January 2016, in Geneva Switzerland[47] extensive and a wide variety of commitment to deliver the Sendai Framework by the science and technology community but it is fragmented. The engagement of the emerging regional Science and Technology Advisory Groups and partners in Africa, Asia, Europe and the Pacific Community will be key to effective partnership building and delivery of the Sendai Framework.

It will be Important to have an effective process for sharing knowledge, building partnerships and clarity on outputs would benefit the S&T programme for Sendai—implementation is critical for the Sendai process, requiring a more joined up approach as science and technology evidence for the Sendai indicators will be important for this process.

---

[44] UN Sustainable Development Solutions Network—Data for Sustainable Development. Available at http://unsdsn.org/what-we-do/thematic-networks/data-for-sustainable-development/. Accessed on 4 August 2015.

[45] Independent Expert Advisory Group on a Data Revolution for Sustainable Development A World that Counts Mobilising the Data Revolution for Sustainable Development Report prepared at the request of the United Nations Secretary-General. 2014. Available at http://www.undatarevolution.org/wp-content/uploads/2014/11/A-World-That-Counts.pdf. Accessed on 4 August 2015.

[46] United Nations E/CN.3/2016/2 Economic and Social Council Report of the Inter-agency and Expert Group on Sustainable Development Goal Indicators. Available at https://unstats.un.org/unsd/statcom/47th-session/documents/2016-2-IAEG-SDGs-E.pdf. Accessed on 4 August 2015.

[47] See Footnote 3.

Annex 1: Global Targets and Indicators of the Sendai Framework Quoted in Full as Adopted by the UN General Assembly in February 2017 (United Nations General Assembly 2017)

	Target A
A-1	Number of deaths and missing persons due to hazardous events per 100,000 populations
A-2	Number of deaths due to hazardous events.
A-3	Number of missing persons due to hazardous events

	Target B
B-1	Degree of direct affectedness by hazardous events per 100,000 population.
B-2 or B-2. alt	Number of injured or ill people due to hazardous events Number of people suffering from physical injuries, trauma or cases of disease requiring medical assistance as a direct result of a hazardous events
B-3a	Number of evacuated people following hazardous events
B-3b	Number of relocated people following hazardous events
B-4	Number of people whose houses were damaged due to hazardous events
B-5	Number of people whose houses were destroyed due to hazardous events
B-6	Number of people who received aid including food and non-food aid due to hazardous events
B-7	Number of people whose livelihoods were disrupted, destroyed or lost due to hazardous events

	Target C
C-1	Direct economic loss due to hazardous events in relation to global gross domestic product
C-2	Direct agricultural loss due to hazardous events
C-3	Direct economic loss due to industrial facilities damaged or destroyed by hazardous events
C-4	Direct economic loss due to commercial facilities damaged or destroyed by hazardous events
C-5	Direct economic loss due to houses damaged by hazardous events
C-5b	Damage and loss of administrative buildings
C-6	Direct economic loss due to houses destroyed by hazardous events
C-7	Direct economic loss due to damage to critical infrastructure caused by hazardous events
C-8	Direct economic loss due to cultural heritage damaged or destroyed by hazardous events
C-9	Direct economic loss due to environment degraded by hazardous events
C-10	Total insured direct losses due to hazardous events

	Target D
D-1	Damage to critical infrastructure due to hazardous events
D-2	Number of health facilities destroyed or damaged by hazardous events
D-3	Number of educational facilities destroyed or damaged by hazardous events
D-4	Number of transportation units and infrastructures destroyed or damaged by hazardous events
D-4b	Kilometres of road destroyed or damaged per hazardous event
D-4c	Number of bridges destroyed/damaged by hazardous event
D-4d	Kilometres of railway destroyed/damaged by hazardous event
D-4 k	Number of airports destroyed/damaged by hazardous event
D-4 l	Number of ports destroyed/damaged by hazardous event
D-1 bis	Number of electricity plants/transmission lines destroyed or damaged by hazardous events
D-5	Number of times basic services have been disrupted due to hazardous events: education (D-5a linked to D-2); water (D-5b linked to D-10)); health (D-5c linked to D-3); sewerage (D-5d); transport (D-5e linked to D-4); government services (D-5f); energy (D-5g); emergency services (D5-h); communications/ICT (D-5i); solid waste (D5-j)
D-14	Number of water and sanitation infrastructures destroyed or damaged by hazardous events

	Target E
E-1	Number of countries that adopt and implement national disaster risk reduction strategies in line with the Sendai Framework for Disaster Risk Reduction 2015–2030
E-2	Percentage of local governments that adopt and implement local disaster risk reduction strategies in line with national strategies

Target F	
Category (a) Financial resources	
Headline indicator for Target F—F-6alt *Two formulations* *The 10 + 10 suggests that a single indicator is selected considering the following 2 options* *It is expected that methodology and data will be further developed over time for ODA and ultimately OOF*	
Option 1	
F-6 alt	Total official international support (ODA plus other official flows) for national DRR actions that is part of **government expenditure**
Option 2	
F-6 alt	Total official international support (ODA plus other official flows) for national DRR actions that is part of a **government-coordinated spending plan**
Supplementary indicators	
F-6a	Total amount of national DRR expenditure

(continued)

(continued)

Target F	
F-6b	Total official international support (ODA plus other official flows) for national DRR actions provided by multilateral institutions
F-6c	Total official international support (ODA plus other official flows) for national DRR actions provided by bilateral entities

	Target G
G-1	Number of countries that have multi-hazard early warning systems
G-2	Number of countries that have a multi-hazard monitoring and forecasting system
G-3	Number of people who have access to early warning information per 100,000 population
G-4	Percentage of local governments having a contingency or emergency plan to act on early warnings
G-6	Percentage of local governments that have multi-hazard risk assessment/risk information, with results in an accessible, understandable and usable format for the people
G-5	Number of countries that have multi-hazard national risk assessment/information, with results in an accessible, understandable and usable format for the people
G-7	Number of people protected per 100,000 population through pre-emptive evacuation following early warnings

References

Aitsi-Selmi A, Murray V (2015) The Sendai framework: disaster risk reduction through a health lens. Bull World Health Organ 93:362. Accessed 3 Aug 2017

Aitsi-Selmi A, Murray V (2016) Protecting the health and well-being of populations from disasters: health and health care in The Sendai Framework for Disaster Risk Reduction 2015–2030. Prehospital Disaster Med. Accessed 3 Aug 2017

Aitsi-Selmi A, Murray V, Wannous C, Dickinson C, Johnston D, Kawasaki A, Stevance A-S, Yeung T et al (2016a) Reflections on a science and technology agenda for 21st century disaster risk reduction. Based on the scientific content of the 2016 UNISDR Science and Technology Conference on the Implementation of the Sendai Framework for Disaster Risk Reduction 2015–2030. Int J Disaster Risk Sci 7(1). https://doi.org/10.1007/s13753-016-0081-x. Available at https://link.springer.com/article/10.1007/s13753-016-0081-x. Accessed on 1 Aug 2017

Aitsi Selmi A, Blanchard K, Murray V (2016b) Ensuring science is useful, usable and used in global disaster risk reduction and sustainable development: a view through the Sendai framework lens. Palgrave Commun 2, Article number: 16016 (2016). https://doi.org/10.1057/palcomms.2016.16. Available at http://www.nature.com/articles/palcomms201616. Accessed 3 Aug 2017

Banwell N, Montoya J, Opeña M, IJsselmuiden C, Law R, Balboa GJ, Rutherford S, Chu C, Murray V (2016) Developing the Philippines as a global hub for disaster risk reduction—a health research initiative as presented at the 10th Philippine National Health Research System Week Celebration. PLOS Curr Disasters 1. https://doi.org/10.1371/currents.dis.5cf90566bb7791456dcf6b9baf6d4873. Accessed 2 July 2017

Deely JD, Ardagh MW (2017) First international conference on people in disasters. Aust J Disaster Trauma Stud 20:57–63 Available at http://trauma.massey.ac.nz/issues/2016-2/AJDTS_20-2_Editorial.pdf. Accessed 31 July 2017

Department of Health (2011) Surveillence in post extreme emergencies and disasters operations manual for managers. Available at http://healthmarketinnovations.org/program/surveillance-post-extreme-emergencies-and-disasters-speed. Accessed 2 July 2017

Dickinson C, Aitsi-Selmi A, Basasbe P, Wannous C, Murray V (2016) Global Community of Disaster Risk Reduction Scientists and Decision makers endorse a science and technology partnership to support the implementation of the Sendai Framework for Disaster Risk Reduction 2015–2030. Int J Disaster Risk Sci 7(1). https://doi.org/10.1007/s13753-016-0080-y. Available at https://link.springer.com/article/10.1007/s13753-016-0080-y. Accessed on 1 Aug 2017

European Commission (2017) Science & policy making—towards a new dialogue. Report on the 2nd International Network for Government Science Advice Conference Science and Policy Making: towards a new dialogue, Brussels, 29–30 September 2016. ISBN 978-92-79-63020-0. https://doi.org/10.2782/977668 HC-06-16-175-EN-N © European Union, 2017 Available at http://www.ingsa.org/wp-content/uploads/2017/03/2nd_ingsa_final_report.pdf. Accessed on 3 Aug 2017

Fakhruddin B, Murray V, Maini R. Disaster loss data in monitoring the implementation of the Sendai Framework ICSU and IRDR. Available at https://www.icsu.org/cms/2017/05/DRR-policy-brief-2-data.pdf (Accessed on 4 August 2015)

Fang C, Shaw R, Abedin MA, Briceno S, Dixit AM, Gupta M, Hugo W, Gensuo J, Murray V, Myint A, Pereira JJ, Rahman A, Sharma VK, Triutomo S, Wang X, Wickramsinghe D (2017) Challenges of disaster risk reduction in the belt and road: contribution of DBAR 52 challenges of disaster risk reduction in the belt and road. Bull Chin Acad Sci 32. Available at http://www.bulletin.cas.cn/publish_article/2017/Z1/2017Z120.htm. Accessed 3 Aug 2017

Gluckman P (2016) The science-policy interface. Science 353(6303):969. Available at http://science.sciencemag.org/content/353/6303/969/tab-pdf. Accessed 3 Aug 2017

Hassan M, ter Meulen V, McGrath P, Fears R (2015) Academies of science as key instruments of science diplomacy. Science and Diplomacy. March 2015. AAAS, Washington DC. Available at http://www.sciencediplomacy.org/files/acadmies_of_science_as_key_instruments_of_science_diplomacy_science__diplomacy_1.pdf. Accessed 3 Aug 2017

Hedlund K (2016) People in disasters conference 2016: learning report—Christchurch earthquake: mental health impacts and psychosocial recovery 2016. ISBN: 978-0-473-39342-7. Available at http://trauma.massey.ac.nz/conference/PID_Learning_Report_2017.pdf. Accessed 31 July 2017

Hemstock S, Buliruarua L-A, Chan E, Chan G, Jacot Des Combes H, Davey P, Farrell P, Griffiths S, Hansen H, Hatch T, Holloway A, Manuella-Morris T, Martin T, Renaud FG, Ronan K, Ryan B, Szarzynski J, Shawl D, Yasukawa S, Yeung T, Murray V (2016) Accredited qualifications for capacity development in disasterrisk reduction and climate change adaptation. Austral J Disaster Trauma Stud 20:16–33 Available at http://trauma.massey.ac.nz/issues/2016-1/AJDTS_20_1_Hemstock.pdf. Accessed 2 Aug 2017

ICSU/ISSC (2015) Review of targets for the sustainable development goals: the science perspective. ICSU/ISSC, Paris. Available at https://www.icsu.org/publications/reports-and-reviews/review-of-targets-for-the-sustainable-development-goals-the-science-perspective-2015. Accessed 3 Aug 2017

International Association of National Public Health Institutes (IANPHI) (2017) Available at http://www.ianphi.org/whoweare/index.html. Accessed 3 August 2017

IPCC (2012) Managing the risks of extreme events and disasters to advance climate change adaptation. In: Field CB, Barros V, Stocker TF, Qin D, Dokken DJ, Ebi KL, Mastrandrea MD, Mach KJ, Plattner GK, Allen SK, Tignor M, Midgley PM (eds) A special report of Working Groups I and II of the Intergovernmental Panel on Climate Change. Cambridge University Press, Cambridge, 582pp. Accessed 2 Aug 2017

Lagmay A (2012) Disseminating near-real time hazards information and flood maps in the Philippines through web-GIS. Project NOAH Open File Reports, 1, 21–36. Available at http://blog.noah.dost.gov.ph/category/open-file-reports/. Accessed 2 July 2017

Liu W, Dunford M (2016) Inclusive globalization: unpacking China's belt and road initiative area development and policy. 1(3). Available at http://rsa.tandfonline.com/action/showCitFormats?doi=10.1080%2F23792949.2016.1232598. Accessed 2 Aug 2017

Maini R, Clarke L, Blanchard K, Murray V (2017) The Sendai Framework for Disaster Risk Reduction and its indicators—where does health fit in? Int J Disaster Risk Sci. https://doi.org/10.1007/s13753-017-0120-2. Available at https://link.springer.com/article/10.1007/s13753-017-0120-2?wt_mc=Internal.Event.1.SEM.ArticleAuthorOnlineFirst. Accessed on 4 Aug 2015

Ministry of Health (2016) Framework for psychosocial support in emergencies. Ministry of Health, Wellington. Available at https://www.health.govt.nz/system/files/documents/publications/framework-psychosocial-support-emergencies-dec16-v2.pdf. Accessed 31 July 2017

O'Mathúna DP, Von Schreeb J (2015) Ethical dilemmas with amputations after earthquakes. In: UNISDR Scientific and Technical Advisory Group Case Studies. Available at http://bit.ly/1z9NVgl

O'Mathúna D, Siriwardhana C (2017) Research ethics and evidence for humanitarian health. Lancet. Available at http://www.sciencedirect.com/science/article/pii/S014067361731276X. Accessed 3 Aug 2017

PHIVOLCS (2016) FaultFinder. Retrieved from http://faultfinder.phivolcs.dost.gov.ph/. Accessed 2 July 2017

Poljanšek K, Marín Ferrer M, De Groeve T, Clark I, Faivre N, Peter D, Quevauviller PK, Boersma KE, Krausmann E, Murray V, Papadopoulos GA, Salamon P, Simmons DC, Wilkinson E, Casajus Valles A, Doherty B, Galliano D (2017a) Science for disaster risk management 2017: knowing better and losing less. Executive Summary. EUR 28034 EN, Publications Office of the European Union, Luxembourg, ISBN 978-92-79-69673-2, https://doi.org/10.2760/451402, JRC102482. Available at http://drmkc.jrc.ec.europa.eu/portals/0/Knowledge/ScienceforDRM/Executive_Summary_of_Science_for_DRM_2017.pdf. Accessed 2 Aug 2017

Poljanšek K, Marin Ferrer M, De Groeve T, Clark I (eds) (2017b) Science for disaster risk management 2017: knowing better and losing less. EUR 28034 EN, Publications Office of the European Union, Luxembourg, 2017, ISBN 978-92-79-60679-3, https://doi.org/10.2788/842809, JRC102482. Available at https://ec.europa.eu/jrc/en/publication/science-disaster-risk-management-2017-knowing-better-and-losing-less. Accessed 2 Aug 2017

Reillon V (2016) Scientific advice for policy-makers in the European Union. European Parliamentary Research Service Briefing. September 2016. European Parliamentary Research Service **Members' Research Service** PE 589.777. Available at http://www.europarl.europa.eu/RegData/etudes/BRIE/2016/589777/EPRS_BRI(2016)589777_EN.pd. Accessed 3 Aug 2017

United Nations (2017) Report of the open-ended intergovernmental expert working group on indicators and terminology relating to disaster risk reduction. United Nations A/71/644 http://www.preventionweb.net/files/50683_oiewgreportenglish.pdf. Accessed 4 Aug 2017

United Nations General Assembly (2017) Report of the open-ended intergovernmental expert working group on indicators and terminology relating to disaster risk reduction United Nations Seventy-first Session A/71/644. Distr.: General 1 December 2016, adopted February 2017. Available at https://www.preventionweb.net/files/50683_oiewgreportenglish.pdf (Accessed on 1 February 2020)

WHO (World Health Organization) (2015) Global reference list of 100 core health indicators. Available at http://apps.who.int/iris/bitstream/10665/173589/1/WHO_HIS_HSI_2015.3_eng.pdf. Accessed 4 Aug 2017

WHO (2016) Guidance for managing ethical issues in infectious disease outbreaks. ISBN 978 92 4 154983 7 © World Health Organization 2016 Available at http://apps.who.int/iris/bitstream/10665/250580/1/9789241549837-eng.pdf. Accessed 3 Aug 2017

Chapter 5
Practical Point of View from Donors: "What We Expect from Science & Technology Group"

Application to Developing Countries

Kimio Takeya and Miki Inaoka

Abstract The Japan International Cooperation Agency supports vulnerable countries affected by natural disasters through disaster risk reduction efforts while contributing to Sendai Framework for Disaster Risk Reduction 2015–2030. At the UNISDR Science and Technology Conference on the Implementation of the Sendai Framework for Disaster Risk Reduction 2015 held at the UN, Geneva in January 2016, Takeya made a presentation at the plenary session on "JICA's Expectations to the Science community." Implementation of the Sendai Framework in most vulnerable areas has always been a problem and Takeya will explore this from a donor's practitioner point of view.

Keywords JICA · DRR · Sendai Framework · Donor · Vulnerable areas · Resilent · Society · Safety level

One of the primary efforts of Japan International Cooperation Agency (JICA) is to support natural disaster prone countries by encouraging disaster risk reduction efforts, and by using the experiences of Japan in line with the Sendai Framework for Disaster Risk Reduction 2015–2030. From a donor's practitioner point of view, Prof. Takeya presented the reality on how to implement the Sendai Framework in the most vulnerable areas, especially in developing countries, and the role disaster risk reduction research to support the activities of donor agencies.

K. Takeya
Japan International Cooperation Agency, IRIDeS (International Research Institute of Disaster Science), Tohoku University, Sendai, Japan

M. Inaoka (✉)
Disaster Risk Reduction Team 1, Global Environment Department, Japan International Cooperation Agency, Tokyo, Japan
e-mail: Inaoka.Miki@jica.go.jp

H. Tatano and A. Collins (eds.), *Proceedings of the 3rd Global Summit of Research Institutes for Disaster Risk Reduction*, Disaster and Risk Research: GADRI Book Series,
https://doi.org/10.1007/978-981-15-8662-0_5

1. JICA's support for disaster risk reduction

1.1 DRR as a development issue

- Poor people live in vulnerable areas, due to their restrictions such as affordable land prices than in the high-end areas. The poorest people often live illegally along the river banks which are originally prohibited as living areas. These communities are more vulnerable to natural hazards. In most cases, they lose their assets, homes and livelihoods, be pushed down to the poorest layer of the society.
- Disaster risk reduction (DRR) is one of the most fundamental responsibilities for the states to secure human lives. The economic development plan that each state tends to be obstructed by natural disasters.
- From these perspectives, DRR must be realized not only as a humanitarian issue but also as a development issue.

1.2 Global DRR support and Japanese position in the Global DRR support

- According to the GFDRR & ODI report: "20 Year Story of International Aid. Financing Disaster Risk Deduction[1]", donor financing heavily depends on Japan and the World Bank. These two entities contribute to more than 55% of the total amount.
- Among the bilateral donors, Japan supports 64% of total assistance, followed by EU with 8%.
- JICA as the implementing agency of Official Development Assistance of Japan has mainly supported "Priority 4: Reduce Risks" of the Hyogo Framework for Action.
- However, according to the above-mentioned report, most of the other financial support was for emergency response & recovery and not towards disaster risk reduction.
- This situation was a big issue. JICA believes that the Global DRR support should shift from reactive support to proactive support, from the humanitarian perspective to development perspective.

1.3 Discussion points at the Post-Hyogo Framework negotiation
JICA as a main member of the negotiation team of the Japanese Government and the Japanese Government proposed and stressed on three important points at the Post-HFA negotiation:
① enhancement of disaster prevention pre-investment.

  - DRR is not a cost but investment, an asset for future development

② empowerment of central governments

  - mainstream DRR into government policies
  - legal, institutional, and budgetary reforming are a must
  - by the leadership of central government, DRR can be mainstreamed into the whole country level

[1] https://www.odi.org/sites/odi.org.uk/files/odi-assets/publications-opinion-files/8574.pdf

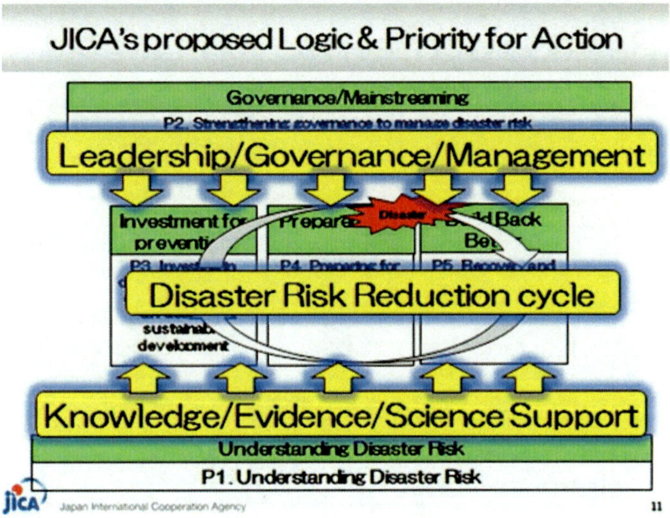

Fig. 5.1 Logical structure of Sendai Framework Priority for Action

③ Build Back Better

- "Build Back Better" concept is to utilize disasters as an opportunity to create more resilient nations and societies than before them through the implementation of well-balanced disaster risk reduction measures, including physical restoration of infrastructure, revitalization of livelihood and economy/industry and the restoration of local culture and environment.
- To prevent future risks, it is indispensable to prevent re-emergence of similar vulnerabilities in the reconstruction processes.
- Support each country's policies, for example, Philippines, Nepal, etc.

2. Logical structure of Sendai Framework Priority for Action (Fig. 5.1).

2.1 Logical sequences of the Priority Actions

- Priority 3 and 4 covers disaster management cycle which is led by governance—Priority 2—and supported by scientific and evidence—Priority 1.
- The structure is shown by the figure below.

2.2 The logical sequences of the seven targets

- Hyogo Framework for Action did not have specific targets, but the Sendai Framework for DRR identifies seven targets.
- The targets consist of three input targets and four outcome targets.
- The four outcome targets show a very important inclusive relationship (Fig. 5.2). Countermeasures to reduce economic losses can also cover the other three targets, whereas countermeasures such as early warning systems that only targets mortality will not be able to reduce other losses.

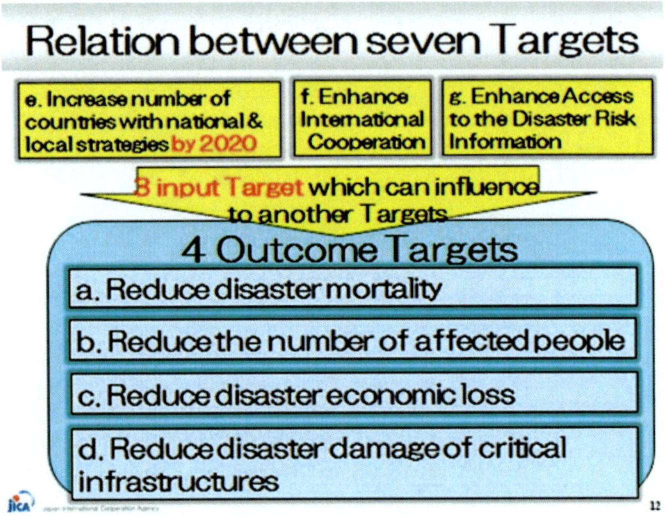

Fig. 5.2 Outcome targets–relation between seven targets

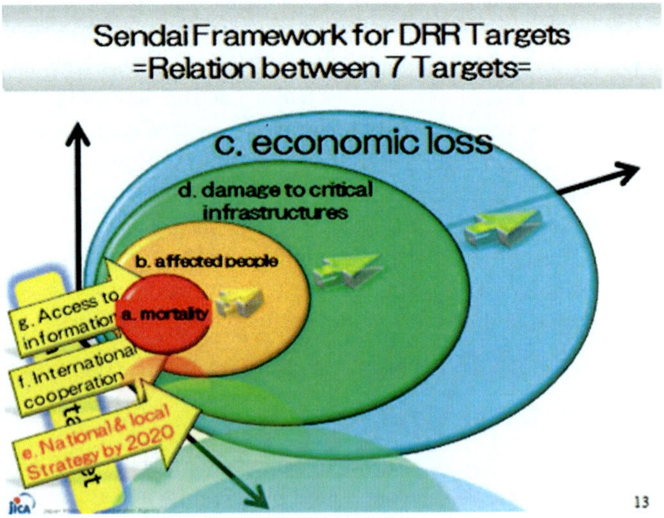

Fig. 5.3 DRR outcome to reduce economic loss

- JICA prioritizes DRR outcome to reduce the economic loss (Fig. 5.3).

2.3 Definition and use of the word "resilient"

- The term "resilient" started to be widely used since HFA, but sometimes developing countries use the word as an excuse to their responsibilities.

- Resilience is defined as "the ability of a system, community or society exposed to hazards to resist, absorb, accommodate to and recover from the effects of a hazard in a timely and efficient manner, including through the preservation and restoration of its essential basic structures and functions" by UNISDR Terminology 2009
- In the case of "School Safety for Earthquake":

  ① if the school is vulnerable, catastrophic collapse will happen immediately:

  - evacuation plan has no sense
  - the students hiding under the desk will not save their lives

  This situation is far away from "resilient".

  ② if the schools are strong and only partially collapsed:

  - evacuation plan makes sense
  - the students hiding under the desk will be effective
  - cushion helmets will be effective
  - but where is the temporary shelter, after survival?

  In this situation, society may be partially (or closer to) "resilient"

  ③ if the schools, hospitals and government buildings are strong enough

  - The facility can function as emergency operation centres
  - The facility can be shelters for refugee

  In this situation, the whole social system may be "resilient"

  - Governments must strengthen school structures. Once this is completed, then they may start with evacuation drills and other post-earthquake training. Many developing country governments tend to put too much focus on evacuation drills when their school is not resilient. This, at the end, causes catastrophic damage and will not allow the children the chance to even evacuate from the school in time.

3. Tailoring to the conditions of each country

   - Figure 5.4 shows the required civil minimum of safety as a function of development phase. The threshold of protected risk and unprotected risk rises as the economy of a country develops.
   - Requirements on civil minimum depend on the matureness of society or level of development. Protection design must be based on the civil minimum and economic conditions of each country (Fig. 5.5).
   - In order to prepare for the uncovered risks, the government will need to mobilize the whole society and all communities to develop capacity and to establish an integrated disaster risk management system.
   - Most of the developing countries still require structural measures and basic infrastructure. Only after they reach to a certain level of investment, then they

Fig. 5.4 Required civil minimum of safety as a function of development phase

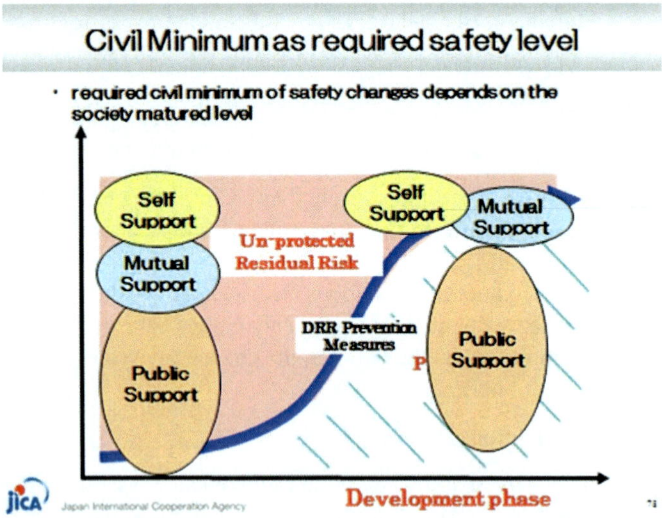

Fig. 5.5 Civil minimum depends on the matureness of society or level of development

will be able to shift to enhancement of mutual-support or self-support portions to respond to residual risks.

4. Gaps to achieve Sendai Framework

   - Target (e) of Sendai Framework is to "substantially increase the number of countries with national and local disaster risk reduction strategies by 2020".

Existing knowledge for each priority area from the practitioner's "brute" view, enough or not?
= not from deliver side but from needs on the ground =

Action Area	Existing Knowledge/Research enough or not? Expected support from S&T
Priority 1	
Understanding disaster risk	hazard/risk mapping ○ **Disaster statistic data stock ×** Standardization of dataset △ **CC/downscaling into practical plan/reduce uncertainty ×** Earthquake pridiction × Using satelite/nondestructive methodology △ Using innovated technology for Meteorology ○
Priority 2	
Strengthening disaster risk governance to manage disaster risk	National/Local DRR Planning ○ **Political mechanism/dynamism research ×** **Convince Political Leader/Social, Political Science ×** Effective Monitaring methodology ×
Priority 3	
Investing in disaster risk reduction for realience	Practical prevention mesures ○ **Macro economic evidence of DRR effectiveness ×** **CC/downscaling into practical plan/reduce uncertainty ×**
Priority 4-1	
Enhancing disaster preparedness for effective response	**Using satelite practical methodology ×** Prevent produce future/underlying risk ×
Priority 4-2	
"Build Back Better" in recovery, rehabilitation and reconstruction	**CC/downscaling into practical plan/reduce uncertainty ×** **Appropriate seismic design for masonry house △**

JICA Japan International Cooperation Agency 31

Fig. 5.6 Table Presented by Prof. Takeya at S&T Conference in Geneva in January 2016

The target year is 2020, ten years earlier than the other targets, to allow time for implementation of the plans and strategies by 2025. Many national strategies have already been prepared based on the Hyogo Framework for Action. However many local strategies still lack reduction of existing risks or prevention of future risks in developing processes.

- Hazard mappings need to be primitive but practical in order to be applicable by local governments. Hazard mappings are critical in enabling local strategies to include appropriate land use plans. Unfortunately, the academia and donors including JICA tend to propose most advanced or sometimes too precise, also too expensive methodologies for hazard mapping. Many developing countries cannot afford to extend these methods.
- Convincing political leaders and economists to invest for DRR remains a challenge. There are very few practical knowledge, evidence or research for political dynamics or economic efficiency of investing not after the disaster but prior to disaster as preventive measures.
- Reducing the uncertainty of future climate change effects is another important issue. Still there are tremendous efforts needed in this area to finalize climate change effect—from global perspective to downscaling to river basin levels for each countries to be able to incorporate the effect in future plans.
- Following table (Fig. 5.6) was presented by Prof. Takeya at the plenary session of the Science & Technology Conference on the Implementation of the Sendai Framework for Disaster Risk Reduction 2015–2030 UNISDR held in Geneva from 27 to 29 January 2016.

Chapter 6
Knowledge to Sustainable Practices: International Network for Transdisciplinary Education (INATE) Approach

Srikantha Herath

Abstract During the past few decades, we have seen great advances in science, technology, information and communications. Yet, real progress is still to be seen in our ability to solve pressing societal problems. The highly complex problems of today spawned by rapid global changes evolve too fast for us to adapt to such changes effectively.

Keywords INATE · Sustainability · Challenges · UNCECAR · Rice terrace systems · Water scarcity · Irrigation

6.1 Introduction

"During the past few decades, we have seen great advances in science, technology, information and communications. Yet, real progress is still to be seen in our ability to solve pressing societal problems. The highly complex problems of today spawned by rapid global changes evolve too fast for us to adapt to such changes effectively. Their inter-connectedness requires a holistic transdisciplinary approach that brings all stakeholders together, including the academe, local government units, private sector, NGOs and communities, to foster rapid feed backs and effective exchange of knowledge and experiences that will enable efficient formulation of sustainable solutions. Transforming conventional project planning to transdisciplinary project design and implementation requires new research on collective solution identification, program design and implementation. This calls for a new type of education and training that promotes transdisciplinary actions. The experiences of a network of leading universities in the region demonstrated that postgraduate sector networks can develop and deliver effective educational and capacity development programs

S. Herath (✉)
International Network for Transdisciplinary Education (INATE) Team Leader,
Center for Urban Water, Ministry of Megapolis and Western Development,
Government of Sri Lanka, Battaramulla, Sri Lanka
e-mail: srikantha.herath@gmail.com

H. Tatano and A. Collins (eds.), *Proceedings of the 3rd Global Summit of Research Institutes for Disaster Risk Reduction*, Disaster and Risk Research: GADRI Book Series,
https://doi.org/10.1007/978-981-15-8662-0_6

to promote such transdisciplinary approaches" (Herath et al. 2015a). This has led to the establishment of International Network for Advancing Transdisciplinary Education (INATE) in 2015 at the World Conference on Disaster Risk Reduction held in Sendai, Japan. INATE promotes transdisciplinary research that act as incubators for development projects. This is done through the concept of **Living Laboratories** that can provide opportunities for incubator research projects implemented in national development programmes that are linked with INATE network. This paper discusses the INATE approach and specific programmes being implemented under in INATE framework.

6.2 Sustainability Challenge

The recent adverse impacts of global environmental change such as climate change, biodiversity loss, global water scarcity has renewed global interest on the need to address human development and global sustainability linkages in a systematic manner. Global environmental concerns grew since late 1960s with the realization of the difficulties associated in trying to meet needs of rapidly growing population with ever increasing demands from earth's limited resources. The UN Earth Summit in Stockholm in 1972 was instrumental in channeling these concerns towards a global movement that demanded regulation and control of impacts of human development activities on nature. "The UN report on development issued by World Commission on Environment and Development, Our Common Future (1987) also known as "Brundtland Report" provided a common platform for different stakeholders and interest groups to discuss ways to address this common goal within each discipline" (Herath et al. 2014a).

6.2.1 Designing Sustainable Systems

"While the definition of sustainable development does not provide a precise mechanism for quantifying sustainability, the flexibility it provided allowed different disciplines to explore its meaning and to communicate across disciplines (Dally Herman 1990). The discussions within each discipline have been converging towards inter-disciplinary approaches on the basis of '*sustainable development* objectives. The major achievement of sustainable development concept is to bring close natural and social sciences (Dally Herman 1990) and its ability to serve as a grand compromise between those who are principally concerned with nature and environment, those who value economic development and those who are dedicated to improving the human condition (Kates et al. 2005). Thus, a sustainable system should be viable and acceptable from environmental, economic and social perspectives. In addition, natural resources base also has ecological functions that keep the earth system as a living organism. This implies maintenance of cyclicity or equilibrium status of major biogeochemical cycles such as carbon cycle, nitrogen cycle and water cycle as well as energy balance of the earth system. Disruptions to these cycles or balances may lead to environmental conditions that are significantly different from the present environment in which the current society has adapted to" (Herath et al. 2014a). Thus it is necessary to consider the complexity, irreversibility, uncertainty associated with earth system processes and ensure that our activities would not threaten the underlying ecological balance. The current approach in designing sustainable development goals, which is based on this approach of ensuring triple bottom line of sustainable development and the sustainability of earth systems processes is a model that can be adopted in designing sustainable systems.

6.2.2 *Ensuring Sustainability of Existing Systems*

Through out history, societies have adapted to changes in environments, governing systems and production methodology changes. Adaptation occurs autonomously when the rate of change is slow. However current rapid global changes such as climate change, urbanization and population increase makes such adaptation difficult. Further, their interconnectedness make them extremely complex to understand and respond, threatening the sustainability of existing systems.

While sustainability is both a desired outcome and a process, the specific actions one need to take has not been clear. In recent times, adapting to change emerged as *Sustainability in Action*. However, forecasting the future we need to adapt is an extremely difficult task. Large uncertainties associated with predictions related to future state of global changes such as climate, makes planned adaptation to an uncertain future a difficult task. Resilience, on the other hand provides us with a positive and pragmatic approach that can be incorporated in regular development activities as well as in adaptation strategies. It allows one to add additional strengths by either catering to multiple sectors or providing diversity. Thus, building resilience is rapidly emerging as the key process to support achieving sustainability.

These concepts are summarised in Fig. 6.1 where subfigure (A) summarises the triple bottom line to sustainable development and sub figure (B) denotes the planetary boundaries. Together they signify the importance of designing systems with sustainable development framework within the bounds of earth system process sustainability. Enhancing resilience will help in ensuring sustainability of existing systems.

6.2.3 *Role of Higher Education and Challenges—UNCECAR*

Adaptation and resilience building to global change is very much a localized activity. It depends on the local hydro-meteorological, geo-physical and socio-economic conditions. Solutions have to be developed locally supported by global knowledge and experiences. For these strategies to evolve locally, local capacity development is

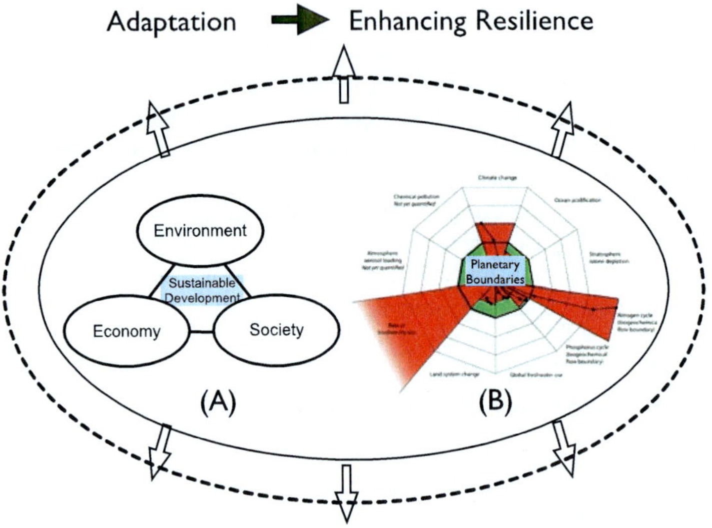

Fig. 6.1 Ensuring sustainability

essential, especially in the developing countries, to customize available global knowledge for local conditions (Herath et al. 2010). This needs to be done in postgraduate education where necessary research can be conducted in applied research projects in partnership with implementing agencies and local communities. Higher education segment of the society has an important role to play in giving direction and leadership in assessing climate change impacts and establishing appropriate frameworks where affected communities can work with specialists in developing appropriate strategies. UNU-ISP organized a regional conference on "The Role of Higher Education in Adapting to Climate Change" in June 2009 at United Nations University headquarters, Tokyo, Japan to discuss ways to take up this challenge. Leading universities in the Asia-Pacific region who gathered at the meeting agreed to establish a University Network for Climate and Ecosystems Change Adaptation Research (UN-CECAR) to collectively develop educational and research programmes (Herath and Surjan 2009).

Research and education are the main focus of the UN-CECAR and the network brings together all available resources and expertise across disciplinary lines to work collaboratively to enhance understanding on climate change impacts and advance adaptation research for the design of appropriate policy and development strategies. Since its inception the universities have jointly developed courses on *Building resilience to climate change, Renewable Energy* and *Leadership for sustainability* by pooling resource and expertise available at 20 leading universities in the Asia Pacific region. Students in all partner universities are eligible to take these courses and transfer credits to their respective postgraduate programs. This approach not only help to share expertise across educational institutions, but also help develop research

Fig. 6.2 Sample activities from the university network for climate and ecosystems adaptation research

teams and researcher networks of both students and faculty on sustainability issues (Herath and Teh 2013). A sample of activities is shown in Fig. 6.2.

6.3 International Network for Transdisciplinary Education (INATE) Approach

At the 3rd World Conference on Disaster Risk Reduction held in 2015 at Sendai, Japan, the University Network for Climate and Ecosystems Change Adaptation Research (UNCECAR) proposed to establish an **International Network for Advancing Transdisciplinary Education (INATE)** to promote transdisciplinary approach to knowledge generation and project implementation by inviting academia, professionals, local governments, private sector and communities to work together to share expertise and experiences. Only through such a network can we secure the best available knowledge, reconcile values and preferences and create ownership of problems as well as solution options, which are all seen as necessary both to solve our real-world problems and to meet our sustainable development objectives. INATE is envisioned to serve as a platform for joint knowledge creation and as an incubator for the translation and implementation of this knowledge to transdisciplinary practices in addressing global change challenges. INATE will conduct a series of research/demonstration projects on a few selected themes to achieve these objectives. Such projects can serve as a catalyst for the development of new interdisciplinary courses and training programmes based on the needs arising from their implementation. In addition, these research/demonstration projects will also address some of the important elements of design and implementation of transdisciplinary projects, such as;

- Identification of critical societal needs to be addressed within the scope of the project.
- Customising global knowledge to local conditions.
- Avoiding conflicts in a multi-disciplinary and multi-stakeholder setting.
- Encouraging active and continuous participation of stakeholders throughout the project duration.
- Maintaining sustainability of transdisciplinary projects.
- Leadership and accountability for implementation of transdisciplinary projects.
- Disseminating and creating adequate awareness among decision takers and planners at local level.

At the end of these research/demonstration projects, it is expected that all stakeholders will have a clear understanding of the challenges, complexities of project implementation and each other's expectations from the outcomes.

The research required for customizing available knowledge and methods to local conditions and confirmation of methodological approaches also can be carried out during this pilot project phase. At the end of the research/demonstration project, this common understanding and the outcomes can be upscaled into development projects by local or national governments. Figure 6.3 describes schematically this concept.

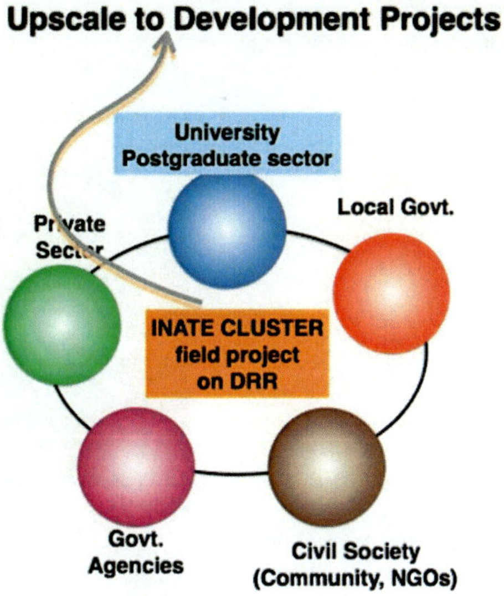

Fig. 6.3 INATE concept of upscaling pilot to development projects

6.4 Case Study 1: Rice Terrace Systems

Rice terrace systems have been developed in many parts of in the mountainous regions of Asia and the Pacific over thousands of years by carving out paddy fields in steep slopes. The Rice Terraces of the Philippine Cordilleras was declared as a UNESCO World Heritage List in 1995 in recognition of its Outstanding Universal Value based on its sustainability, communal management, and its testimony to the harmonious relationship between people and their environment.

The Hani rice terraces located in the Yuanyang County of the Yunnan province in China have been known for their sustainable resource utilization methodologies for over 1000 years. This system also contains the cultural landscape of the Honghe Hani Rice terraces, and was declared as a UNESCO world cultural heritage site in 2013 (Herath et al. 2015b, c).

These systems have evolved through the interactions between society and nature. Their sustained existence is a proof of the fact that development is more robust when the measures taken are in harmony with the natural surroundings. However the special ecological functions of rice terraces are very much dependent on complex water management and are vulnerable to climate and environment change, as well as development pressures which threatens their sustainability.

A comparative research study was undertaken to assess sustainable development strategies for Hani Rice terraces Ifugao rice terraces in the Philippines under the framework of INATE as a research project of UNU-IAS postgraduate programme with funding from Asia Pacific Network for Global Change Research (APN) (Herath et al. 2015d).

Fig. 6.4 Rice terraces in Hani, Yunan province, China

Fig. 6.5 Rice terraces in Hani, dried up conditions in 2010

Hani rice terraces, are shown in Fig. 6.4 while the dried up conditions which led to this research is shown in Fig. 6.5.

The rice terrace systems research project initially focused on the analysis of stability of rice terrace systems under future climate change impacts. The objective was to analyze vulnerability due to increase of both flood and drought extremes under future climate scenarios and propose ecosystems based approaches, especially augmentation of water retention structures in the head waters of the study sites located in the Ifugao rice terrace systems in the Philippines and the Hani rice terrace system in China. However consultation with local stakeholders revealed the need for a much broader holistic approach for the sustainable development of the rice terrace systems considering not only ecological resilience but also societal and economic resilience.

A framework was developed to address the diverse drivers and understand their interconnectedness by the author. This framework was used to address building

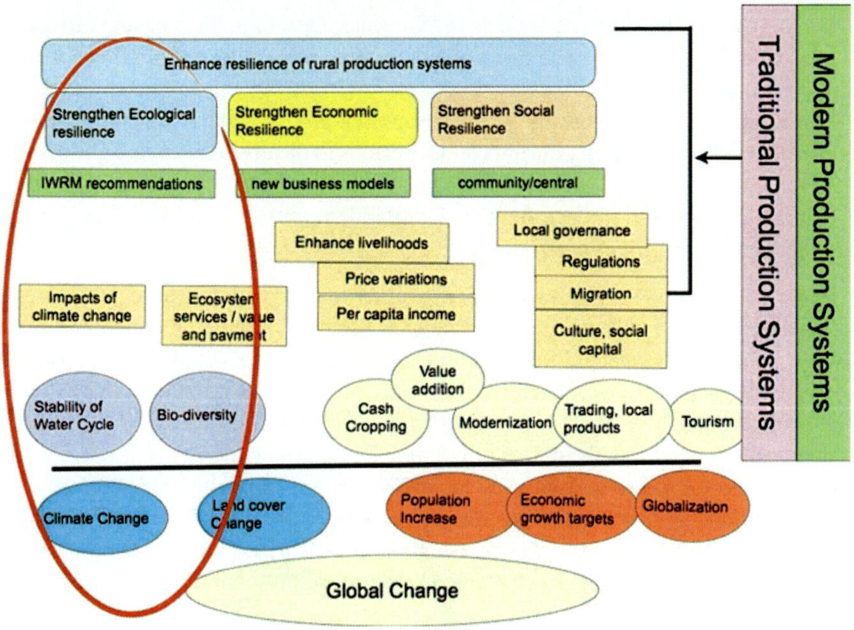

Fig. 6.6 Framework adopted to address resilience

resilience from Ecological, Economic and Social perspectives. The framework is shown in Fig. 6.6.

Thus, the research expanded to provide a rich venue for the development of trans-disciplinary research case studies for the post graduate programme of UNU-IAS. Four groups of students conducted field studies over two years to develop research proposals in consultation with local communities, academics, NGO and government agencies. At the end of the project a public forum was organized in Manila, Philippines to share experiences and exchange views on future development pathways for the two rice terrace systems.

Sustained and sustainable existence requires a thorough understanding and assessment of the entire system, the existing linkages and associated complexities, particularly in relation to the natural resources, water being one of them. In designing development strategies, it become necessary to focus on an overarching objective that is of common interest to all sectors of the community and broad enough to link with different development issues to analyze and resolve trade-off issues.

For the Hani rice terraces in China, a development strategy centered around water security was proposed for integrating development programmes to promote equitable and sustainable development. Some of the key findings of the study related to Hani Rice Terraces in China are,

1. Groundwater is a critical component of the water system in the area and its interplay with the surface flow and storage determines the water availability for the downstream region.
2. The River-ditches network in the upstream area provides stable irrigation water but there are water scarcity during certain periods in the upstream areas. The effect of scarcity is being felt more in the downstream reaches of the system.
3. Climate change will impact the water availability, with dry and wet seasons set to become more pronounced.

Water Scarcity Index (Rws) which is the ratio of water withdrawal to availability was selected as the index to characterise water security in three upstream and two downstream villages selected for the study. As reliable water withdrawal data were not available, water use estimated from field survey and consumption patterns was used in the calculation of Rws. $Rws \leqslant 0.1$ means no stress, where as a Mild to moderate stress corresponds to Rws of 0.2–0.3. Bias corrected CORDEX (EC-Earth model) climate projections were used to analyse water availability under future climatic conditions. Future demands were estimated for expected tourism demand, crop diversification and population change. Change of present and future Water Scarcity index shows increased stress in Feb and March, more pronounced for downstream than upstream.

As a counter measure doubling of water retention capacity of upstream ponds was considered. Figure 6.7 shows a snapshot of ponds located upstream and the Fig. 6.8 shows seasonal variation of Rws at present, for future changed climate conditions and future state with doubling of upstream retention ponds as a counter measure (shown in orange colour).

Some of the main recommendations made were

1. Development of a holistic, basin level strategy should be done for collective decision making.
2. Establishment of a nodal authority for improving inter-agency coordination along with development of local platforms for engaging, involving and consulting local stakeholders has to be done.
3. The water scarcity index is useful in showcasing the system threats and determining the timing and effect of on-ground interventions to aid effective decision making.

In the Ifugao terraces in the Philippines, a pilot study site has been established in Bayninan, Kiangan with an automatic rain gauge, surface flow measurement and ground water level observation. The Ifugao rice terraces and the threats due to high discharges are shown in Fig. 6.9a, b.

Projections for future climate were obtained from bias-corrected outputs of the MRI-GCM and HadGEM3-RA RCM. Of the simulations done, results show that the wet seasons of 2021–2030 pose high risk of excess runoff and slope instability. Appropriate measures such as improvements in the drainage system must be implemented to address these risks. Meanwhile, the simulations also show that the

Fig. 6.7 Upstream Ponds in Hani Rice terraces

dry seasons at the end of the century (2091–2100) are at high risk of water deficits, particularly in the months of April, May, and June.

To address excess runoff, the spillways in the paddies must be expanded to allow the release of large volumes and avoid the collapse of the terrace walls. Also, these months present an opportunity to store excess water for use during the dry months. Construction of small reservoirs upstream in the forests is a possibility if viable mechanism can be found to combine water storage and forest management. In the Nagacadan rice terrace cluster of the Ifugao Rice Terraces a study was carried out to assess the carbon stock of local forests by measuring four carbon pools of the local forests; the forest carbon stock, non-tree vegetation, forest floor litter layer, and soil carbon 10 cm depth. Two types of forest that are native to the area were used in the carbon study, the *muyong* and the *bilid* together with measurements in open lands and abandoned paddy. The *muyong*, refers to the private woodlotsthat are maintained by residents of the area. The *bilid* is the other forest land use in the area.

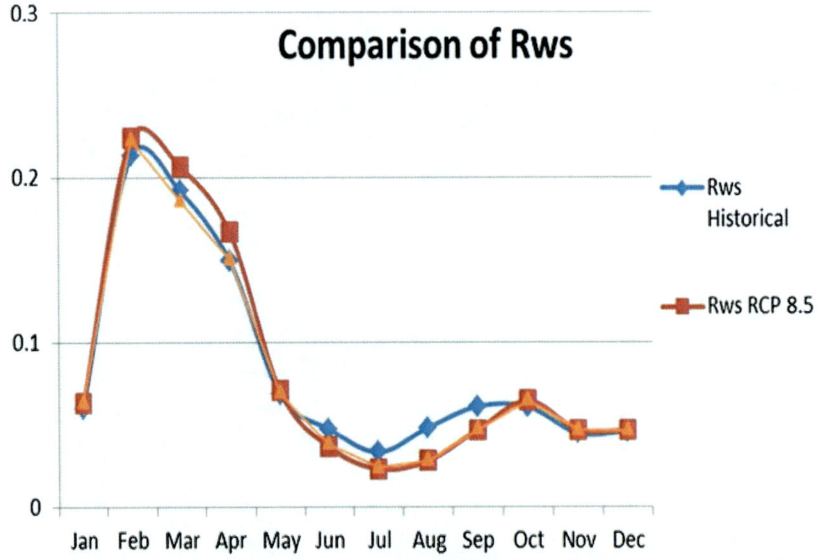

Fig. 6.8 Water Scarcity Index, present, future and with ponds

(a) Ifugao Rice Terraces, Philippines (b) Paddy field bund erosion from high discharges

Fig. 6.9 Ifugao rice terraces and threats

Field measurements were carried out to estimate carbon storage in these different vegetation groups and are summarised in Fig. 6.10. The study showed clearly that muyong forests with traditional farming practices has a much higher carbon sequestration capacity compared to other types of landuse. Therefore an integrated approach through traditional forest management supported by REDD+ funding is proposed for livelihood enhancement through water management. Early warning of slope stability and rainfall forecasts systems installed in the study site can be used to continuously enhance local capacity for climate information use and improve understanding of the ecological functions of the terrace system (Herath et al. 2015e, f, g, h).

6.5 Case Study 2: Mosaics in Irrigation

"There are many traditional agricultural production systems in Asia that have resulted not only in outstanding landscapes, maintenance of agricultural biodiversity, indigenous knowledge and resilient ecosystems development but also provided economic, environmental and social goods and services over thousands of years. With growing population and economic aspirations, many of these systems are being replaced by modern agriculture systems that are designed for efficiency and large-scale rapid development. However, there is also a growing realization that we should in some form preserve these valuable repositories of indigenous knowledge for climate change adaptation, biodiversity conservation and land management and the rich culture they spawned. Different approaches such as Cultural Heritage Systems of UNESCO, or the Globally Important Agriculture Heritage Systems of FAO attempt to preserve and showcase representative production sites from these systems. However, they cannot be upscaled to cover the vast populations still engaged in them. In this research we investigated the feasibility of fusing the traditional and the modern systems as shown in Fig. 6.11 through building mosaics of traditional and new systems" (Herath et al. 2018a)

The Deduru Oya reservoir, which was commissioned in 2014, is primarily planned to improve the livelihood of farmers in parts of the North-Western Province of Sri Lanka by increasing the productivity of its land and water resources by regulating and diverting water to irrigation systems through two main canals along both river banks.

The left bank canal supplement water needs for paddy cultivations from existing ancient rain-fed small reservoir based irrigation systems. The right bank canal is a

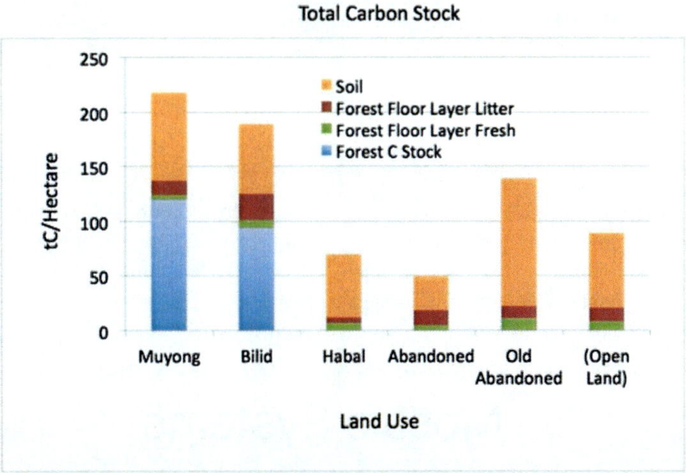

Fig. 6.10 Carbon storage of various forest types in Ifugao

trans-basin canal conveying excess water from the reservoir to adjacent Mee oya basin. The project location and the system description is shown in Fig. 6.12.

The Deduru Oya irrigation project as a Mosaic System will supplement the ancient irrigation tanks through the left bank feeder canal. A water allocation model using Water Evaluation and Planning (WEAP) tool developed by the Stockholm Environmental Institute was used to analyse the optimal water utilization of 134 ancient irrigation systems fed by the main reservoir. Inflows to main reservoir as well as each individual reservoir was simulated separately with hydrological models and the whole system combined with irrigation demand for two season cultivation was analysed using the WEAP model.

The simulation carried out using rainfall for past ten years reveal that most of the ancient irrigation tanks alone cannot meet the two season irrigation water requirements. The new Deduru Oya Reservoir storage in normal climatic year is well enough to meet all the irrigation, hydropower and environmental flow requirements. However, the total irrigation demand cannot be accommodated by the new reservoir alone at extremely dry weather at 20% monthly inflows. If combined with the ancient irrigation systems, there can be sufficient water to meet irrigation demands provided the whole system is managed with prioritising water availability for the month of September. These simulation results are shown in Fig. 6.13.

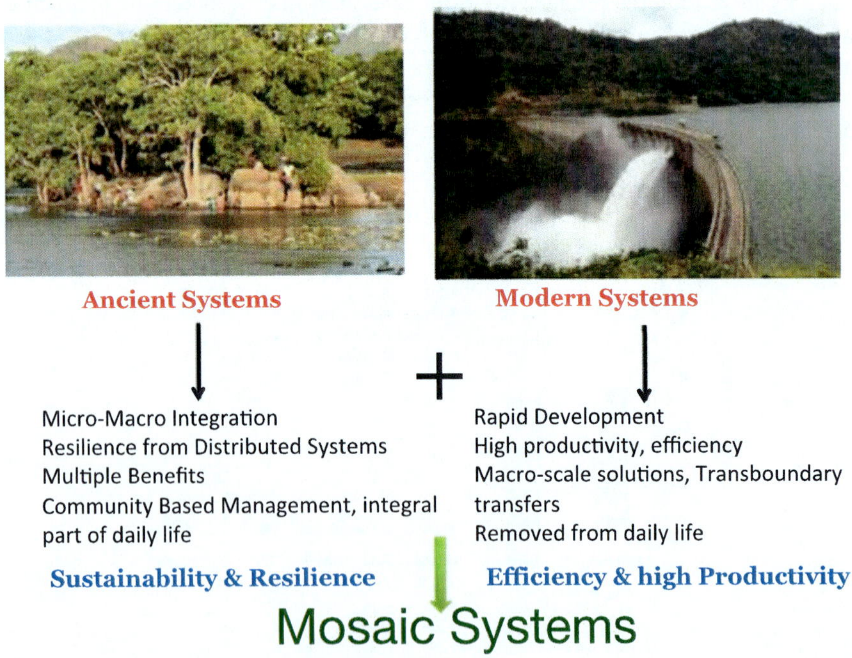

Fig. 6.11 Moscaics from ancient and modern systems

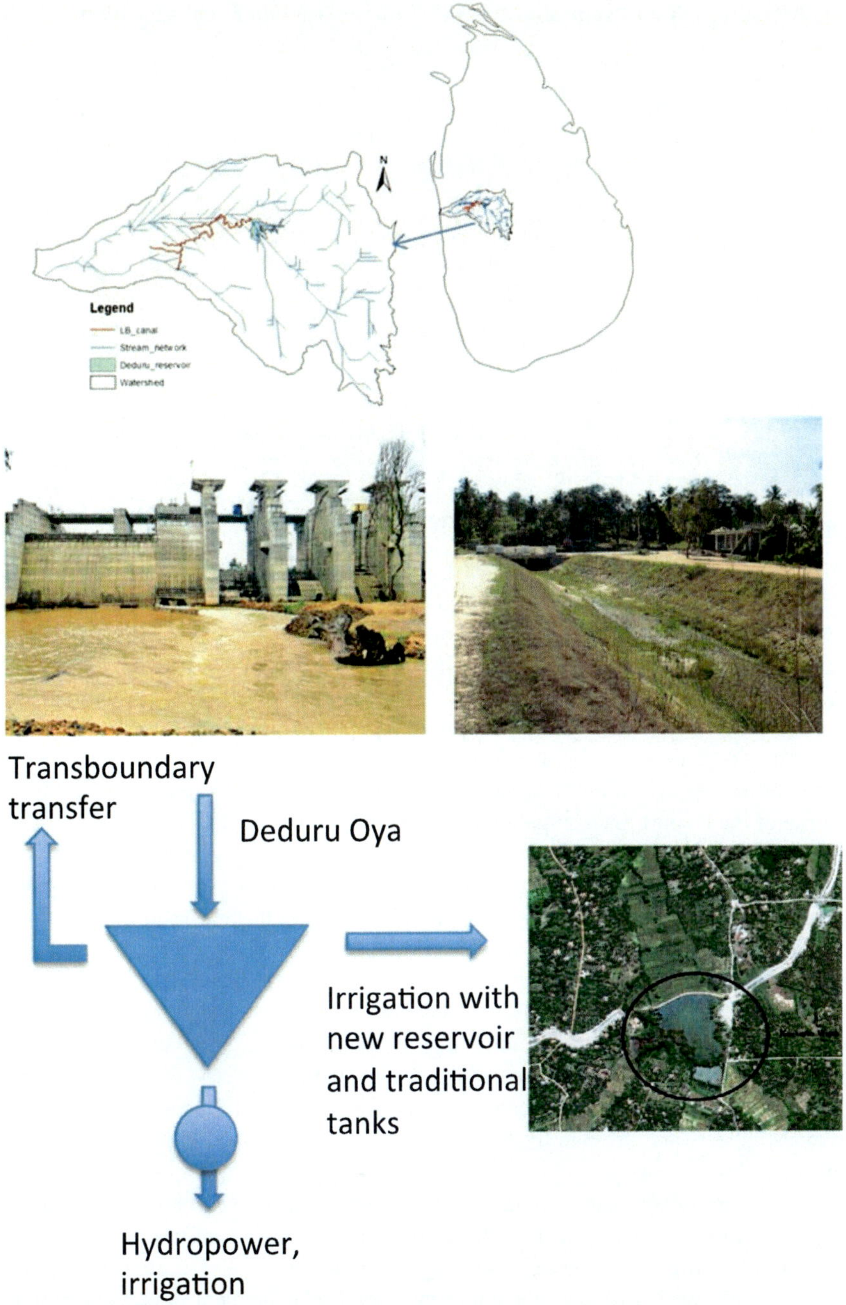

Fig. 6.12 Deduru Oya system

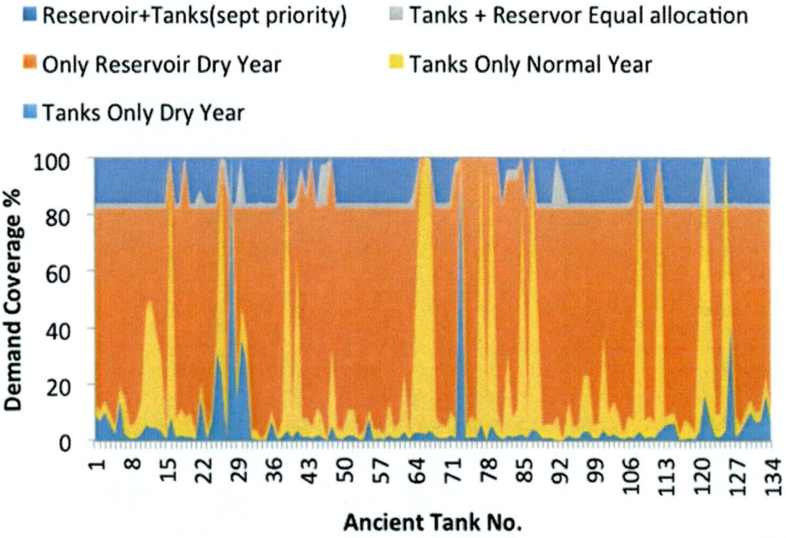

Fig. 6.13 Deduru Oya System performance for various configurations

The use of ancient irrigation systems is not only to provide resilience for extreme dry conditions, but also to provide the social and cultural cohesion and harmony with nature. The enhanced water resources provide opportunities to improve livelihoods of the farmers in the region. Thus, a holistic approach towards empowering the farmer communities through integrated water management practices is necessary to make full use of the Deduru Oya irrigation project.

The challenge facing this approach is the design of an appropriate water allocation system and implementing a robust water management system. The Bulk Water Allocation (BWA) model adopted in system H of the Mahaweli basin in the North Central Province of Sri Lanka was studied and found to be a promising model to be used in the Deduru Oya basin project. The assessment of implementing BWA through a detailed survey of the Deduru Oya basin has been carried out and showed a positive attitude by the farmer community. In implementing the BWA, it is important to establish (a) clear and measurable water entitlement, (b) incorporate 'risk management' in comprehensive capacity building that includes social, technical and financial aspects and provide (c) transparency in decision-making and appropriate power sharing.

The Deduru Oya project provides a unique opportunity to combine the efficient large-scale modern systems with resilient localized ancient systems promoting social and harmony with nature though building mosaic systems. These mosaics should cover both physical (structural mosaics) and social (management mosaics) aspects. Further analysis on economic aspects, in linking micro production with macro economy through crop and livelihood diversification should be studied in future. The

(a) Kelani River Basin and Parliament under floods in 2010

(b) Drainage system of Metro Colombo area

Fig. 6.14 Metro Colombo drainage system and urban flooding

present study clearly identify the importance of transdisciplinary approach in carrying out such complex projects involving multi-stake holder engagement.

6.6 Case Study 3: Managing Urban Water: A Living Laboratory for INATE

"Flooding in Colombo metropolitan area has for many years been hampering public life and obstructing traffic and commercial activities on a regular basis. Figure 6.14a shows the location of the city and the flooding of national parliament in year 2010. Figure 6.14 shows the macro drainage system of the city which has 4 outfalls to river and 3 outfalls to sea that has a low discharge capacity due to low surface gradients. The flooding is due three major causes: (a) lack of outfall capacity (b) lack of storage capacity, and (c) bottlenecks in the river and canal system (bridges, culverts, narrow river bends, blocking bridge piers/abutments, pipeline crossings, etc.).

In 2012 the World Bank and the Government of Sri Lanka (GoSL) agreed upon a loan for a Metro Colombo Urban Development Project (MCUDP). The overall Project Development Objective of the Loan is to support the GoSL in:

1. reducing flooding in the catchment of the Colombo Water Basin
2. strengthening the capacity of local authorities in the Colombo Metropolitan area to rehabilitate, improve and maintain local infrastructure and services through selected demonstration investments.

The flood control projects are divided in to two groups; one primarily aiming at the macro-drainage system, including the greater Colombo catchment/basin, and one aiming at the micro-drainage system, which primarily covers the metropolitan area. Measures include new diversions (canals or tunnels), widening of river/canal stretches, river-bend widening, smoothening or bend cut-offs, enlarging existing outfalls, construction of new gate structures, upgrading street drains, enlarging culverts, etc.

In addition, apart from structural measures a number of non-structural measures are being considered to strengthen the flood management capacity, of which the most important component is a Real-Time Control (RTC) System, and/or a Flood Early Warning System (FEWS). In a related development, in 2015, the Government of Sri Lanka announced an ambitious plan to develop the Western Province to a large Megalopolis, and a new ministry, the Ministry of Megapolis and Western Development (MMWD), was established to implement the programme. Sri Lanka Land Reclamation and Development corporation that implement the Metro Colombo Development Project also comes under MMWD. For the Ministry of Megapolis and Western Development assessing short term and long-term flood risks is important to safeguard its investment and ensure sustainable urban development. Thus it was decided to enhance the scope of the proposed RTC to incorporate also current and future risk assessment. A new facility to handle Flood Control and Water Management is currently being setup that will

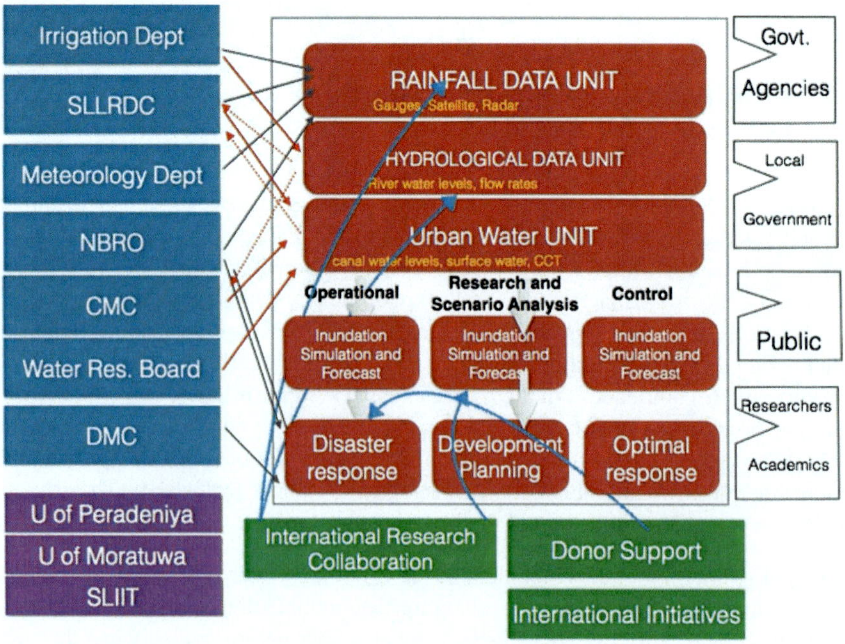

Fig. 6.15 The framework for center for flood control and water management

1. Develop an integrated flood control and water management information system
2. Provide early warning support for the Metro colombo area
3. Develop optimal operational rules for the flood control facilities such as pumps and storage facilities considering also the potential storage and use water to make a pleasant urban environment.
4. Assess current and evolving future risk to Megapolis from urban development as well as climate change.

The author is at present engaged in setting up this center in Colombo, Sri Lanka. Sustainability of the System is a major consideration that is being addressed in the development stage. A sustained capacity development programme with the collaboration of educational and research institutions in the country, support for continuous R&D, fostering collaboration among of academia and the line agency professionals are some of the key concerns addressed together with the development of the center" (Herath 2017). Figure 6.15 shows the system design and this collaborative framework. The center currently has arrangements with two leading engineering universities so that the work carried out by its staff can be the thesis requirement for Master degree. The staff members in this programme spend one working day of a week at the university as well as Saturdays and Sundays attending lectures to complete a one year research Master degree while being employed by the center.

Maintenance of the System in the future require diversification of its activities beyond disaster management. Integrating urban water management with flood control, Catering to Megapolis development needs; Risk Assessment, Cost benefit analysis are some of the areas the centre the center is currently engaged with. The Megapolis project, through the flood control and water management center will provide opportunities for postgraduate researchers of INATE network to be engaged in real-life applied research projects as a living laboratory.

6.7 Conclusions

Implementing sustainable development programes faces many challenges due to rapid global changes. Planning specific measures as well as Investment under uncertain future conditions is extremely difficult, especially for developing countries. Building redundancy, diversity, resistance and recovery mechanisms with multi-stakeholder involvement is a pragmatic approach to reduce adverse impacts of global change.

Transdisciplinary approaches provide best opportunities for implementing sustainable programmes that are compatible with user needs and are localised with local physical and social characteristics. Rather than conventional top down approaches, we need to take an iterative bottom up approach that is designed and implemented with the user community. INATE proposes an approach where a pilot project is first launched that can absorb the risks and develop a robust, sustainable solution to the problem being addressed. Higher Education Sector can play a major role in bringing up these transformations.

References

Dally Herman E (1990) Sustainable development: from concept and theory to operational principles. Population Dev Rev 16:25–45

Herath S (2014) Advancing a nexus approach to the sustainable management of water, soil and waste. Reference Material for the Dresden Nexus Conference 2015 (White Book). United Nations University Institute for Integrated Management of Material Fluxes and of Resources (UNU-FLORES)

Herath S (2017) Keynote address: Urban Water Management in Metro Colombo. In: Engineering Research Conference (MERCon), Moratuwa. IEEE, New York, pp xxi-xxii

Herath S, Mishra B, Wong P, Weerakoon SB (2018) Mosaic of traditional and modern agriculture systems for enhancing resilience. Resilient Asia. Springer 2018:151–187

Herath S, Diwa J, Yuanmei J, Castro PPM (2015b) Overview of rice terrace farming systems in Hani and Ifugao: Water Management and Current Threats, UNU-IAS working paper series

Herath S, Jayaraman A, Diwa J (2015c) Ensuring water security for the sustainability of the Hani Rice Terraces, China against Climate and Land Use changes, UNU-IAS working paper series

Herath S, Jayaraman A, Diwa J (2015f) Study of institutional and governance arrangements for achieving water security in the Hani Rice Terraces, UNU-IAS working paper series

Herath S, Kawasaki J (eds) (2012) Comparative studies on development strategies considering impacts of adaptation to climate change, CECAR series No. 12, UNU-ISP, p 265

Herath S, Kenichi T, J Diwa (2015e) Assessment on the feasibility of REDD+ in Nagacadan Rice Terraces of Ifugao and its Muyong forest, UNU-IAS working paper series

Herath S, Lim SA, Wang Y, Takahashi S (eds) (2012) Higher Education for climate and ecosystems change adaptation Asia, Traditional Knowledge, CECAR series No. 9, UNU-ISP, p 89

Herath S, Lim A, Takahashi S (Eds) (2010) Working with local communities - higher education for climate and ecosystems change adaptation: Asia. Proceedings of the 3rd UN-CECAR conference and workshop, The Role of Higher Education in Adapting to Climate and Ecosystems Change

Herath S, Mario M, Diwa J (2015g) Bias-corrected daily precipitation estimates in the Ifugao Rice Terraces under Climate Change Scenarios, UNU-IAS working paper series

Herath S, Mario M, Diwa J, Beredith B (2015d) Surface and groundwater flow response to climatic change in the Ifugao Rice Terraces, UNU-IAS working paper series

Herath S, Mohri H, Diwa J (2015a) University Led Development: Proceedings of 8th UN-CECAR International Conference and Workshop, CECAR Series 17, UNU-IAS

Herath S, Surjan A (Eds) (2009) Higher Education for Climate and Ecosystems Change Adaptation: Asia. Proceedings of conference and workshop, the Fist Consultation Conference on The Role of Higher Education in Adapting to Climate and Ecosystems Change, Tokyo, Japan, 10-12 June 2009

Herath S, Teh SH (eds) (2013) UNCECAR postgraduate courses on Renewable Energy - Handbook 2012, CECAR series No. 13, UNU-ISP, p 83

Herath S, Teh SH, Mohri H, Mahalingam A (eds) (2012) Higher education for climate and ecosystems change adaptation Asia, Disaster Risk Management and SustainabilityÓ, CECAR series No. 10, UNU-ISP

Herath S, Tsusaka K, Diwa J (2015h) Assessment on the feasibility of REDD+ in Nagacadan Rice Terraces of Ifugao and its Muyong forest. Rice Terrace Farming Systems Working Paper Series. United Nations University Institute for the Advanced Study of Sustainability

Kates RW, Parris Thomas M, Leiserowitz Anthony A (2005) What is sustainable development? Environment 47(3):8–21

Chapter 7
Disaster Risk Management Knowledge Centre: A New European Initiative to Bridge Science and Policy

Tom De Groeve, Montserrat Marin Ferrer, Ian Clark, and Karmen Poljanšek

Abstract As the science and knowledge service of the European Commission, the Joint Research Centre's mission is to support EU policies with independent evidence throughout the whole policy cycle. Faced with the risk of increasingly severe and frequent natural and man-made disasters, policy-makers and risk managers in disaster risk management (DR).

Keywords DRMKC · European initiative · Science and policy · EU · DRM · Partnerships · Knowledge · Innovation

As the science and knowledge service of the European Commission, the Joint Research Centre's mission is to support EU policies with independent evidence throughout the whole policy cycle. Faced with the risk of increasingly severe and frequent natural and man-made disasters, policy-makers and risk managers in disaster risk management (DRM) and across EU policies increasingly rely on the wealth of existing knowledge and evidence at all levels—local, national, European and global—and at all stages of the DRM cycle—prevention; reduction; preparedness; response and recovery. Better knowledge, stronger evidence and a greater focus on transformative processes and innovation are essential to improve our understanding of disaster risk, to build resilience and risk-informed approaches to policy making, and contribute to smart, sustainable and inclusive growth. To tackle the challenges of fragmented knowledge and fragmented networks, the JRC has created the Disaster Risk Management Knowledge Centre. The Disaster Risk Management Knowledge Centre (DRMKC) provides a networked approach to the science-policy interface in DRM, across the Commission, EU Member States and the DRM community within and beyond the EU. This paper discusses early results in three parts: partnerships and networks to improve science-based services; better use and uptake of research and operational knowledge; and innovative tools and practices for risk and crisis management.

T. De Groeve (✉) · M. Marin Ferrer · I. Clark · K. Poljanšek
European Commission, Joint Research Centre (JRC), CCR - TP267, Via Fermi 2749, 21027 Ispra, VA, Italy
e-mail: tom.de-groeve@ec.europa.eu

H. Tatano and A. Collins (eds.), *Proceedings of the 3rd Global Summit of Research Institutes for Disaster Risk Reduction*, Disaster and Risk Research: GADRI Book Series,
https://doi.org/10.1007/978-981-15-8662-0_7

7.1 Introduction

Population growth, climate change, globalisation and interconnected systems of trade, finance and information create new risks. Policymakers and risk managers in disaster risk management (DRM) must face the risk of increasingly severe and frequent natural and man-made disasters, with impacts cascading through the economy, environment and society. In this complex setting, there are increasing calls for an evidence-based approach. The new EU legislation (Decision (EU) 2019/420) explicitly calls to "increase the availability and use of scientific knowledge on disasters" and "take action to improve the knowledge base on disaster risks, and to better facilitate and promote cooperation and the sharing of knowledge, the results of scientific research and innovation, best practices and information, including among Member States that share common risks". Acknowledging that there is a wealth of existing knowledge and evidence at all levels—local, national, European and global—and at all stages of the DRM cycle—prevention; reduction; preparedness; response and recovery, the legislation calls for "a Union Civil Protection Knowledge Network [...] taking due account of the expertise available in the Member States and the organisations active on the ground".

At a global level, the Sendai Framework for Disaster Risk Reduction shifts the emphasis from response-oriented disaster management to comprehensive disaster risk management, in which a more systematic and reinforced science-policy interface strengthens the contribution of DRM to smart, sustainable and inclusive growth globally (Murray et al. 2015).

Two key challenges in the scientific world are increased complexity and acceleration (Poljansek et al. 2017). Ever more science is produced and is available at a mouse-click. Ever more actors from different disciplines and policy areas are involved. For practitioners, policy makers and even for scientists themselves, the challenge now is to find the relevant science, from multiple disciplines, and make sense of it, for multiple policies.

In the inherently interdisciplinary field of disaster risk management, this is very evident (Poljansek et al. 2017). The science is complex due to the interaction of vast and varied communities on Earth sciences, and their interaction with the engineering communities, social and behavioural scientists and economists. Equally complex is the policy side (Bower et al. 2017). Disaster risk is linked to civil protection, climate change, environmental, health, humanitarian, and development policy, but also to fiscal and economic policy directly. A holistic approach is necessary.

In order to improve all stages of the DRM cycle, the knowledge and evidence base need to be further improved, advances in relevant technology extensively exploited and research results broadly applied, while promoting enhanced interaction between researchers and end users. Sharing knowledge across the DRM cycle and relevant stakeholders is a key point of this process. This need is echoed in the Science and Technology Roadmap for Sendai Framework, developed at the 2016 UNISDR Science and Technology Conference.[1]

[1] https://www.unisdr.org/partners/academia-research/conference/2016/.

7.2 Disaster Risk Management Knowledge Centre (DRMKC)

The Disaster Risk Management Knowledge Centre[2] is a collaboration of scientists, policy makers and practitioners, and aims to be a nexus where knowledge exchange is facilitated. It was initiated by the Joint Research Centre (JRC) of the European Commission. The JRC is the European Commission's science and knowledge service which employs scientists to carry out research in order to provide independent scientific advice and support to EU policy. By being both a scientific organisation and part of the European Institutions, it is a "boundary organisation" that facilitate collaboration and information flow between diverse research disciplines and between the research and public policy community. In 2015, the JRC started to emphasise its role as a boundary organisation more explicitly, including through the creation of Knowledge Centres in specific, multi-disciplinary fields (JRC 2016).

The Disaster Risk Management Knowledge Centre, or DRMKC, was the natural outcome of an increased collaboration between scientists in JRC and policy officers in various Directorates General of the European Commission, including in Civil Protection and Humanitarian Aid Operations, Climate Action, International Cooperation and Development, Migration and Home Affairs, Health and Food Safety, Environment and Research and Innovation. As science provides evidence to various policy areas, it can—if communicated and interpreted in a context-specific way—become a common ground for cross-sectoral dialogues. In the European Commission, disaster risk falls mainly under the Civil Protection legislation, but is clearly linked to policies in 11 Directorates General (Bower et al. 2017). Also disaster risk reduction outside of Europe spans across individual competences and the European Commission's Action Plan on the Sendai Framework for Disaster Risk Reduction 2015–2030 (European Commission 2016) groups actions across policy areas in a coherent way.

The role of science in disaster risk reduction has been recognised by policy makers. The 2013 Civil Protection legislation (European Union 2013) addresses disaster risk in both the EU and in third countries. The new legislation of 2019 (Decision (EU) 2019/420) calls for strengthening the evidence base and sharing knowledge. A similar emphasis on the role of science in disaster risk reduction was included in the Sendai Framework for Disaster Risk Reduction (UNGA 2015). The DRMKC was launched in October 2015 to respond to both European and global needs. In fact, at the UNISDR Science and Technology Conference on the Implementation of the Sendai Framework for Disaster Risk Reduction, the European Commission communicated that the Disaster Risk Management Knowledge Centre was its contribution to the Science and Technology Roadmap.

The DRMKC organises its work in three pillars (partnership, knowledge and innovation) with each 2 objectives (Fig. 7.1). It revises its work plan on an annual basis with input from scientists and policy makers in EU Member States, participating

[2]https://drmkc.jrc.ec.europa.eu.

Fig. 7.1 Pillars and objectives of the Disaster Risk Management Knowledge Centre

Directorates General and JRC scientists. Some key actions are described in the next sections.

7.3 Partnership

To achieve the ambitious goal of fully exploiting and translating complex science into useful policy and applications in DRM, the DRMKC reinforces the development of disaster science partnerships and networks. In the EU, Member States have implemented a variety of mechanisms for science-policy interfaces (De Groeve and Casajus 2015; Marin Ferrer et al. 2016). The UK's Natural Hazard Partnership,[3] Italy's Centres of Competence[4] in the Civil Protection Department and France's Disaster Observatory[5] are but three examples. At European level, the DRMKC focuses on two types of partnerships: those for providing scientific advice during emergencies, and those for providing scientific advice for policy making.

[3]https://www.naturalhazardspartnership.org.uk (accessed 8/8/2017).

[4]https://www.protezionecivile.gov.it/jcms/en/centri_competenza.wp?request_locale=en (accessed 8/8/2017).

[5]https://www.onrn.fr/ (accessed 8/8/2017).

First, partnerships for operational preparedness and response to major natural disaster types in the EU are promoted to facilitate the information flow between the different partnerships, the Emergency Response Coordination Centre (ERCC) and Member States. The ERCC is the operational heart of the EU Civil Protection Mechanism. It plays a key role as a coordination hub to facilitate a coherent European response during emergencies inside and outside Europe. Scientific advice during emergencies is provided partly by the Joint Research Centre directly and partly in a new partnership called ARISTOTLE.

The JRC (2015) provides information through its European early warning systems (European Flood Awareness System, European Forest Fire Information System, and European Drought Observatory), which are all being expanded to global coverage. In addition, the Global Disaster Alert and Coordination System (GDACS) provides key scientific information for all hazards to the wider humanitarian community, the Epidemic Intelligence from Open Sources (EIOS) system serves the World Health Organisation, and eMars collects lessons learnt for chemical accident prevention. The JRC also implements the Copernicus Emergency Management Service[6] (Wania et al. 2016) which provides maps based on earth observation for response, reconstruction and risk assessment globally. Each system has its own networks of scientists, practitioners and policy makers, and the DRMKC fostered cross-sectoral and cross-disciplinary linkages to enable mutual learning and data sharing. The Knowledge Centre also explores models with distributed competence centres. ARISTOTLE[7] is a pilot project, implemented through a consortium of research centres from EU Member States, that aims at providing accurate and authoritative information on natural disaster to ERCC through three cornerstones: designing a flexible multi-hazard scalable early warning service, comprehensive reporting of the available information gathered by the 24/7 operational centres, providing expert analysis by an expert panel readily available upon request through, e.g. teleconferencing.

Second, networks and activities are activated and promoted to improve the science-policy interface in prevention activities and to facilitate the translation of complex science into useful policy advice. Currently, the DRMKC groups 11 networks and ensures information exchange among them. The network topics range from multi-hazard risk assessment methodologies to recording disaster loss data, from space technology to critical infrastructure protection, from Natech to floods and space weather issues. There is specific emphasis to link disaster risk and climate change adaptation communities, e.g. through collaboration with the PLACARD[8] project

[6]https://emergency.copernicus.eu (accessed 8/8/2017).

[7]https://aristotle.ingv.it/ (accessed 8/8/2017). ARISTOTLE is a pilot project that aims at providing accurate and authoritative information on natural disaster.

[8]https://www.placard-network.eu/ (accessed 8/8/2017). PLACARD's (PLAtform for Climate Adaptation and Risk reDuction) mission is to be the recognised platform for dialogue, knowledge exchange and collaboration between the Climate Change Adaptation (CCA) and Disaster Risk Reduction (DRR) communities.

and the Climate ADAPT[9] platform. Also the nexus disaster risk and space technology is prominent, e.g. through participation in Copernicus and the Group on Earth Observation.

The DRMKC uses various tools to stimulate information exchange among disciplines, networks and sectors, including bi-monthly newsletters, joint events, co-authorship in publications and an Annual Seminar.

7.4 Knowledge

Scientific research results and operational knowledge gained from lessons learnt, exercises, training, peer reviews and other assessment tools need to be better exploited in the DRM cycle to mitigate risks and vulnerabilities and to improve response when disaster strikes.

To this end, the Disaster Risk Management Knowledge Centre has produced first flagship science report "Science for disaster risk management 2017: knowing better and losing less" (Poljansek et al. 2017). A second report is in preparation to be published in 2021. It gathers views and contributions from nearly 300 scientists, policy makers and practitioners. Like the first summary, it will support the integration of science into informed decision making through synthesising and translating evidence for disaster risk management and strengthening the science-policy and science-operation interface.

In a multi-disciplinary field like disaster risk management, it is critical to share knowledge and practice across sectors and domains. In support of national risk assessment exercises (mandatory in the EU under the Union Civil Protection Mechanism legislation), the DRMKC is coordinating a process to develop science-based recommendations for disaster risk assessments across all hazards. This fosters a holistic understanding of both disaster risk—addressing hazards, exposure and vulnerability—as well as disaster risk management—focusing on the four phases of disaster cycle: prevention, preparedness and response, reconstruction, and risk financing—through a systematic multi-hazard assessment overview of existing disaster risk knowledge.

The knowledge work of the DRMKC highlights the further need of new research avenues to address the multi-risk impacts of natural and human-induced hazards as well as the increasingly systemic nature of risk (GAR 2019).

To exploit the results of the many research projects funded by European research programmes, the European Commission has set up a Community of Users, where scientists and users frequently meet to discuss results, gaps and future research needs. In this context, the DRMKC curates a repository of all relevant research and operational projects and published gap analyses to inform the global and EU research agendas (Fig. 7.2).

[9]https://climate-adapt.eea.europa.eu/. The European Climate Adaptation Platform (Climate-ADAPT) aims to support Europe in adapting to climate change.

Fig. 7.2 Overview of the DRMKC project explorer, featuring 512 projects with participation of 2546 organisations. https://drmkc.jrc.ec.europa.eu

7.5 Innovation

Transferring knowledge into practice requires innovation. The participation of industry, and particular the small and medium enterprises, is essential to modernise disaster risk management practices. Operationalisation of research, such as in the Copernicus Emergency Management Service (Wania et al. 2016), directly facilitates the work of first responders and other operational actors in crisis management through innovative technologies and instruments.

The DRMK's innovation pillar is the European Crisis Management Laboratory (ECML). The ECML acts as a research, development and test facility for ICT focused solutions. It also is a back-up facility for the operational emergency room of the

European Commission. This dual role creates a virtuous feedback loop between research and operations that drives innovation. In particular, the laboratory specialises in the integration of devices, applications and crisis management-related information sources to support crisis management needs, such as threats analysis, common situation awareness and collaborative decision making.

7.6 Conclusions

In its five years of existence, the Disaster Risk Management Knowledge Centre (DRMKC) has contributed to better disaster risk management in the EU. It has broadened the knowledge base, built awareness across policy, scientific, sectoral and geographic boundaries, forged new partnerships and fostered innovation. The DRMKC underpins the work of relevant Commission services, Member States and the wider DRM community, and represents a coherent contribution to the Sendai Framework for Disaster Risk Reduction. The DRMKC is a regional approach for creating an efficient science-policy interface, taking account of national specificities and mandates. To share science solutions and policy challenges beyond the region, the Disaster Risk Management Knowledge Centre is open to reach out to similar initiatives in other regions and globally.

References

Bower A, Blok J, Dali M, Faivre N, Fell T, Ghislain P, Happaerts S, Kavvadas I, Kockerols P, Molnar AM, Quevauviller P, Villette F (2017) Current status of disaster risk management and policy framework. In: Poljanšek K, Marín Ferrer M, De Groeve T, Clark I (eds) Science for disaster risk management 2017: knowing better and losing less. EUR 28034 EN, Publications Office of the European Union, Luxembourg, Chapter 1. https://doi.org/10.2788/688605

De Groeve T, Casajus Valles A (2015) Science policy interfaces in disaster risk management in the EU. Mapping the support provided by science in the EU civil protection mechanism. EUR 27520. Publications Office of the European Union, Luxembourg (Luxembourg)

European Commission (2016) Commission staff working document: action plan on the Sendai framework for disaster risk reduction 2015–2030: a disaster risk-informed approach for all EU policies. Publications Office of the European Union, Luxembourg, SWD (2016) 205 final/2

European Union (2013) Decision No 1313/2013/EU of the European parliament and the council of 17 December 2013 on a union civil protection mechanism. Official J Eur Union L 347:924–947

JRC (2015) Science for disaster risk reduction and response. https://ec.europa.eu/jrc/en/publication/brochures-leaflets/science-disaster-risk-reduction-and-response. Accessed on 11 Apr 2019

JRC (2016) Annual report 2015. Publications Office of the European Union, 2016, JRC 98031, p 32. https://doi.org/10.2788/209

Marín Ferrer M, Casajus Valles A, Civiletti G (2016) Science policy interfaces in disaster risk management in the EU: requirements and conditions for efficient SPIs in practice. EUR 28280 EN. Publications Office of the European Union, Luxembourg (Luxembourg). https://doi.org/10.2788/196166

Murray V, Aitsi-Selmi A, Blanchard K (2015) The role of public health within the United Nations post-2015 framework for disaster risk reduction. Int J Disaster Risk Sci 6(1):28–37

Poljanšek K, Marín Ferrer M, De Groeve T, Clark I (2017) Science for disaster risk management 2017: knowing better and losing less. EUR 28034 EN, Publications Office of the European Union, Luxembourg, ISBN 978-92-79-60678-6. https://doi.org/10.2788/688605, JRC102482

UNGA (2015) Resolution adopted by the General Assembly on 3 June 2015, 69/283. Sendai framework for disaster risk reduction 2015–2030. United Nations General Assembly, 23.6.2016, A/CONF.226/4

Wania A, Broglia M, Kucera J, Spruyt P, Rossi M, Spinoni J, Weissteiner C, Dorati C, Steel A, Louvrier C (2016) User workshop of the copernicus emergency management service—summary report; ISBN 978-92-79-63450-5. https://doi.org/10.2788/2672

Part III
Selected Papers from Presentations (Arranged in Country Alphabetical Order)

Chapter 8
International Federation for Information Processing and Its Domain Committee on Information Technology in Disaster Risk Reduction

Dimiter Velev and Plamena Zlateva

Abstract The chapter describes the origin, goals and the main figures of Domain Committee on Information Technology in Disaster Risk Reduction (DCITDRR), established by the International Federation for Information Processing (IFIP). Key activity, such as international conferences organized by DCITDRR, and some of the scientific research projects of DCITDRR members are presented.

Keywords Disaster · Risk · Reduction · Information Technology

8.1 Introduction

The effects of natural disasters are very serious and the destruction caused may take a very long time to recover. Related damages are severe and it may cause relief expenses of billions of Euros. With the increase of natural disasters that have occurred in the past years, it is expected their frequency will continue to increase in the coming years (UNISDR EUR 2016). In this respect, it is important to point out that on 18 March 2015 at the Third UN World Conference on Disaster Risk Reduction in Sendai City, Miyagi Prefecture, Japan, the UN Member States adopted the so-called Sendai Framework for Disaster Risk Reduction 2015–2030 (UNISDR 2015). It aims for the following outcome: "The substantial reduction of disaster risk and losses in lives, livelihoods and health and in the economic, physical, social, cultural and environmental assets of persons, businesses, communities and countries." At present, at global and national levels, a wide range of multiple scientific and scientific-applied research activity in the area of disaster risk reduction concerning individual types of disasters is conducted (UNISDR 2017).

Modern information and communication technologies can facilitate significantly the decision-making processes from the point of view of disaster risk reduction. Due to disaster risk reductions multidisciplinary nature, people from various backgrounds, such as industry, diverse geographical and global settings, not-for-profit organization

D. Velev (✉) · P. Zlateva
University of National and World Economy, Sofia, Bulgaria
e-mail: dgvelev@unwe.bg

H. Tatano and A. Collins (eds.), *Proceedings of the 3rd Global Summit of Research Institutes for Disaster Risk Reduction*, Disaster and Risk Research: GADRI Book Series,
https://doi.org/10.1007/978-981-15-8662-0_8

backgrounds, agriculture, marine life, welfare, risk management, safety engineering and social networking services will be included.

8.2 Presentation of the International Federation for Information Processing

One of the organizations that have recently focused its attention on the problem of disaster risk reduction is the International Federation for Information Processing (IFIP, https://ifip.org). Formally, IFIP is a non-governmental, non-profit umbrella organization for national societies working in the field of information processing. It was established in 1960 under the auspices of UNESCO as a result of the first World Computer Congress held in Paris in 1959. IFIP is incorporated in Austria by decree of the Austrian Foreign Ministry (20th September 1996, GZ 1055.170/120-I.2/96) granting IFIP the legal status of a non-governmental international organization under the Austrian Law on the Granting of Privileges to Non-Governmental International Organizations (Federal Law Gazette 1992/174).

The International Federation for Information Processing:

- is the leading multinational, apolitical organization in information and communications technologies and sciences;
- is recognized by United Nations and other world bodies;
- has a formal consultative partnership relation with UNESCO;
- represents IT societies from 56 countries or regions, covering all five continents with a total membership of over 500 000 members;
- sponsors 100 conferences yearly from theoretical informatics to the relationship between informatics and society including hardware and software technologies, and networked information systems.
- links more than 3500 scientists from academia and industry, organized in more than 101 Working Groups (WG) reporting to 13 Technical Committees (TC):
  - TC 1: Foundations of Computer Science
  - TC 2: Software: Theory and Practice
  - TC 3: Education
  - TC 5: Information Technology Applications
  - TC 6: Communication Systems
  - TC 7: System Modeling and Optimization
  - TC 8: Information Systems
  - TC 9: ICT and Society
  - TC 10: Computer Systems Technology
  - TC 11: Security in Information Processing
  - TC 12: Artificial Intelligence
  - TC 13: Human–Computer Interaction
  - TC 14: Entertainment Computing.

Within IFIP, the IT societies' members find a meeting place for sharing experience, and discussing challenges and opportunities, as well as the TCs and WGs contribute to, and often lead, progress in the state-of-the-art knowledge and practice.

The IFIP scientific and technological leadership is warranted by WG membership, based solely on individual excellence and it is asserted by the organization of some 100 highest quality international events, and the publication of some 30 new books annually that are distributed worldwide.

8.3 Establishment of the Domain Committee on Information Technology in Disaster Risk Reduction

Following the increasing number of disasters worldwide and its growing ITC potential and expertise, IFIP at its General Assembly, held on 8–9 October 2015 at the Daejeon Convention Center, Daejeon, Korea, has established the **Domain Committee on Information Technology in Disaster Risk Reduction** (DCITDRR).

The main goals of the DCITDRR are to:

- promote disaster risk reduction (DRR) within the IT community;
- provide an additional opportunity for IFIP members to work with external specialized bodies such as UN, UNISDR, ICSU, ITU and ISCRAM;
- coordinate the efforts of member societies as well as different Technical Committees and Working Groups of IFIP in its specific disaster-related field.
- The DCITDRR activities are based on the following major pillars:
- Information acquisition and provision:
  - People search: safety information: online, cell phone, offline;
  - Visualizing lifeline information: road condition, transport, electricity, water supply, etc.;
  - Radioactivity, chemical pollution, etc.;
  - Websites of disaster information;
  - Networking for information infrastructure: Internetworking with communication links.
- Shelter information management for a local government:
  - List of people in a shelter: name/age/family/address;
  - Information systems for food and goods distribution;
  - Volunteer support.
- Disaster Information Systems:
  - Information systems which are different from a normal-time use;
  - Need for a standard format—safety information; information on suffer: family, shelter; good distribution: never be well-planned; medical information: the disaster weak; donation: traceability;

  - Open source and global community of software developers as Sahana and Ushahide;
  - Global collaboration over the net;
  - Need well-known interfaces.

- Top of the art information and communication technologies:

  - Social media;
  - Mobile computing;
  - Internet of things;
  - Big data;
  - Cloud computing;
  - Artificial intelligence.

The DCITDRR has the following management board:

- Chair: Prof. Dr. Yuko Murayama, IFIP Vice-President, Japan.
- Vice Chairs:

  - Diane Whitehouse, IFIP TC9: ICIT and Society chair, UK;
  - Assoc. Prof. Dr. Plamena Zlateva, Bulgarian Academy Sciences, Bulgaria;
  - Prof. Dr. Erich Neuhold, IFIP TC5 Information Technology Application former chair, Austria;

- Secretary: Prof. Dr. Dimiter Velev, TC5 New Activities and Interdisciplinary Research chair, Bulgaria;
- IFIP Liaison Secretary: Eduard Dundler, Austria.

DCITDRR has presented its activity to the world for the first time at the UNISDR Science and Technology Conference on the Implementation of the Sendai Framework for Disaster Risk Reduction 2015–2030 in Geneva, Switzerland, 27–29 January 2016, followed by related activities at other workshops, conferences and events. The latest presentation is at the Third Global Summit of Research Institutes for Disaster Risk Reduction—Expanding the Platform for Bridging Science and Policy Making (GSRIDRR 2017), Kyoto University, Kyoto, Japan, 19–21 March 2017. All these presentations focus on the role, expertise and potential of the IFIP and its DCITDRR to collaborate with researchers and scientists from all over the world to elaborate complex and efficient ICT systems in fighting disasters.

8.4 DCITDRR Conferences

Meanwhile DCITDRR has conducted its First IFIP Conference on Information Technology in Disaster Risk Reduction (ITDRR-2016), followed by the preparation of its second one (ITDRR-2017), https://itdrr.unwe.bg/, as an attempt not only to popularize its existence, but also to provide an open forum for the dissemination of the current research progress, innovative approaches and original research results on the application of Information Technology in Disaster Risk Reduction. The conference focuses on important topics, such as:

- Advanced ICT and natural disasters
- Big data and natural disasters
- Climate change and disaster risk
- Climate information management
- Climate information processing
- Cloud computing in emergency management
- Communications and natural disasters
- Crowdsourcing and emergency management
- Disaster information processing
- Disaster prevention and mitigation
- Disaster relief, resilience and research
- Disaster risk management capability assessment
- Emergency preparedness
- Hazard, vulnerability and risk mapping
- ICT challenges in emergency management
- Integrated risk governance
- IoT and natural disasters
- Mobile computing and emergency management
- Natural disaster monitoring
- Natural disaster risk assessment
- Natural disaster risk management
- Natural disaster risk reduction
- Natural disasters and security
- Natural hazard and vulnerability analysis
- Security and privacy issues in disaster management
- Social media and natural disasters
- Socio-economic impacts of natural disasters, etc.

8.5 Some R&D Projects by DCITDRR Members

Currently, Prof. D. Velev, the DCITDRR secretary, who is also the Director of the Science Research Center for Disaster Risk Reduction at the University of National and World Economy, Sofia, Bulgaria, manages three disaster-related projects:

1. Information System for Integrated Risk Assessment from Natural Disasters, Grant No. DFNI-I02/15, 2014–2018, funded by the Bulgarian National Science Fund. In accordance with the national and European priorities for the security, the purpose of the project is to develop a general framework of information system for integrated risk assessment from natural disasters with the help of modern information and communication technologies.

The information system will include innovative tools for an integrated risk assessment from natural disasters, which will be implemented on the basis of the developed new and adapted existing mathematical methods.

The information system will be designed on the principle of modules, which will implement interaction with geographical information system (GIS), and heterogeneous databases containing information about the objects and for potential natural disasters on Bulgarian territory.

The information system will use cloud computing and specialized online social network for the exchange of heterogeneous databases and expert knowledge for the risks of natural disasters.

The information system will assist those interested in risk assessment for test sites before, during or after the occurrence of natural disasters.

The information system can successfully serve professionals from different fields of science as a single platform for interdisciplinary research regarding the risk assessment from natural disasters.

The information system could be used as a part of the overall systems for the prevention and management of emergency situations.

The information system will assist the responsible representatives (State, municipal, etc.) to make informed decisions regarding the efficient allocation of limited financial resources (public and private) for the prevention and management of risk between the investigated objects.

The following research tasks are formulated in order to achieve the project's aim:

- A systematic study of the problem area: determination of potential sources of risk (natural disasters, possible for Bulgarian territory) and analysis of the mutual influence between them.
- Examination and analysis of existing approaches, methods and models to assess the risk of a variety of natural disasters;
- Analysis and selection of appropriate modern information technologies (Cloud computing, big data, online social networks and others, which together with heterogeneous databases and GIS to be included in a single information system;
- Proposing approaches and mathematical models for integrated risk assessment from natural disasters based on developed under the project new and/or adapted existing methods;
- The development of a general framework of information system, unifying the proposed models and selected information technologies for an integrated risk assessment from natural disasters;
- Approbation of the general framework of the information system with examples of specific disasters and locations, verification of the results obtained and the formulation of requirements for the implementation of the system;
- Dissemination of the results among all interested in the topic and developing specialized modules for training in risk assessment from natural disasters.

The project implements an innovation approach for risk management of natural disasters based on ISO 31,000:2009 "Risk management—principles and guidelines."

The innovation integrated risk assessment approach includes following major activities that are designed as elements of the information system: 1. Comprehensive description of the monitored objects; 2. Risk identification—description of the possible disasters and assessment of the vulnerability of the monitored objects; 3. Risk analysis of the monitored objects; 4. Risk evaluation of the monitored objects.

The proposed integrated information system should be regarded as a Web-based research software platform, which includes the following individual and logically connected components:

- Subsystem for risk source identification; risk component analysis, economic assessment of consequences; complex risk analysis and evaluations;
- Subsystem for integrating various data sources in GIS environment through developed models;
- Dedicated social network site (SNS) for helping the monitoring and control of emergency situations.

The project idea is to provide software tools for modeling and analysis of the monitored objects as a result of miscellaneous natural hazards, an assessment of the mutual impact and negative consequences, as well as the provision of a complex information for the facilitating the process of making managerial decisions. The SNS should solve common problems and implement features, which are needed to obtain and store data from/to different physical locations, to verify and filter the data, to avoid information overload, to use ready models for disaster management, to enable collaboration, to provide for reliability and accessibility, to attract experts and users in particular emergency managers.

2. A Conceptual Model of Cloud-Based Information System for Risk Assessment from Natural Disasters, Grant No. DNTS/China 01/6, 2014–2018, a Bulgarian—Chinese R&D Project (University of National and World Economy, Sofia, Bulgaria—Shenyang University of Chemical Technology, Shenyang, China).

The main objective of the project is to develop a conceptual model of a Cloud-based information system (CBIS), which to implement the proposed innovative methodology for integrated risk assessment of natural disasters in Bulgaria and China. The CBIS should serve as a unified platform for interdisciplinary research of the risk using cloud computing. The following aims for implementing the project's main objective are defined to be achieved:

- Definition and analysis of risk sources (possible disasters) and risk components (monitored objects), as well as their interrelationship for both countries—Bulgaria and China;
- Economic and financial assessment of the consequences (losses) for the monitored objects;
- Complex risk assessment by the use of classical and intelligent mathematical methods;
- New information retrieval for the levels of extremity of the monitored objects and presentation of the corresponding recommendations for efficient distribution

of the financial funds, defined for preventing and reducing risk from possible disasters;
- Integration of heterogeneous databases with chronological and expert information for the possible disasters for both countries—Bulgaria and China;
- Development of new and adaptation of appropriate existing methods for risk analysis and evaluation;
- Organization and conducting of thematic conferences and training seminars in both countries—Bulgaria and China.
- Development of educational modules for the Ms.Sc. courses and special training.

The risk management of natural disasters is usually performed in conditions of the subjectivity and incomplete certainty. This requires application of innovative methods for risk analysis/assessment and development of sophisticated CBIS for emergency management.

The results obtained by project will be disseminated in Bulgaria and Chine. Developed website and prepared information materials about the project results in English, Bulgarian and Chinese.

3. Research on the Applicability of Virtual Reality in Education and Business, Grant No. 22/2017, 2017–2019, funded by the University of National and World Economy, Bulgaria.

The main objective of the project is to analyze the specificities and make recommendations for the implementation of virtual reality (VR) in disaster risk reduction training. The idea is to justify the use of VR to simulate multiple scenarios regarding the occurrence of different types of disasters. These scenarios are planned to be used to train professionals with different profiles to increase their capacity for disaster risk reduction and effective emergency response.

VR scenarios can also be used to assess the vulnerability of different sites in specific geographic regions to specific disasters.

The proposed VR scenarios will allow changing the input parameters (different types and severity of disasters), tracking the relevant outputs (negative effects on different sites) and analyzing possible actions of trainees in specific emergencies.

The project's expected results are that VR implementations will increase the quality and effectiveness of disaster risk reduction training.

8.6 Conclusions

Currently, the DCITDRR lists more than 20 members from different interdisciplinary areas who actively work and contribute to promoting disaster risk reduction within the IT community and they coordinate the efforts of member societies as well as different IFIP Technical Committees and Working Groups in specific disaster-related fields. DCITDRR organizes workshops and the ITDRR conference to establish an academic environment that fosters the dialogue and the exchange of ideas between different levels of academic, research, business and public communities.

The Domain Committee on Information Technology in Disaster Risk Reduction at the International Federation for Information Processing represents an open forum for discussing local and world problems due to disasters, for sharing the results of current research progress in disaster risk reduction, for working out common goals and projects, which will ultimately help society successfully fight disasters.

References

United Nations Office for Disaster Risk Reduction (UNISDR) (2015) Sendai framework for disaster risk reduction. https://www.unisdr.org/we/coordinate/sendai-framework

United Nations Office for Disaster Risk Reduction—Regional Office for Europe (UNISDR EUR) (2016) European forum for disaster risk reduction (EFDRR) Roadmap 2015–2020. https://www.unisdr.org/we/inform/publications/55096

United Nations Office for Disaster Risk Reduction (UNISDR) (2017) Cross-sectoral and multi-risk approach to cascading disasters. https://www.unisdr.org/we/inform/publications/57080

Chapter 9
Research Capacity in Disaster Risk Reduction—An Indian Perspective

Jaya Singh and Ajay Gairola

Abstract The world is continuously experiencing natural hazards associated with threat of loss of lives and damage to property. At a large scale, many communities are strongly engaged in disaster risk reduction DRR activities; however, it is still a challenge how to integrate groups of scientific and technical networks for obtaining, sharing, and implementing collected data during the existing disasters, and based on these results, a policy making and implementation process can be generated. The 3rd Global Summit of Research Institutes for Disaster Risk Reduction (GSRIDRR-2017) expands the knowledge of natural or man-made disaster risks and enhances the scientific and technical research efforts as well as the key actions for policy roadmaps for DRR. For this, Global Alliance of Disaster Risk Institutes (GADRI) is contributed an important role to further facilitate the contributions of research institutes, local bodies and organizations for disaster risk reduction for plan a sustainable future.

Keywords Research capacity · DRR · India perspective · Sustainable · Challenges · Flood risk

9.1 Introduction

Natural disasters are complex and collective events resulting from lack of risk management strategies that reflect the factors of historical and current conditions (Alexander 2000). It is noticed that in many countries there has been an increase in the risk of natural disasters due to environmental degradation (World Bank 2002). International strategy for disaster reduction in Geneva 2001 defines a "disaster" as a serious disturbance to the society and widespread human, material or environmental

J. Singh (✉)
National Centre for Medium Range Weather Forecasting, Noida sector 62, India
e-mail: Jayasingh2525@gmail.com

A. Gairola
Indian Institute of Technology, Centre of Excellence in Disaster Mitigation and Management, Roorkee, Uttarakhand, India

H. Tatano and A. Collins (eds.), *Proceedings of the 3rd Global Summit of Research Institutes for Disaster Risk Reduction*, Disaster and Risk Research: GADRI Book Series,
https://doi.org/10.1007/978-981-15-8662-0_9

losses. To achieve the target for disaster risk reduction, the contribution research institutes, organizations (central, provincial, local, and ministries), stakeholders, agencies as well as local communities will need to work together and be aware of their respective roles and responsibilities. By this effort, a policy-making process can be established for DRR.

An effective disaster risk reduction covers the following—

(1) **Risk identification**—It is necessary to identify and understand the problem which can include the current hazards maps, household's vulnerability, risk modeling, and capacity analysis.
(2) **Prevention**—It includes all possible measures to avoid the hazardous phenomena turn into disaster for well establishment of society, its economies, and infrastructure.
(3) **Risk mitigation**—It has a wide range of preparedness and pre-disaster recovery planning which include structural design, land-use planning, construction practices, building codes, public awareness, early warning systems, and preparing response plan.
(4) **Response**—It is taken immediately prior to and following the disaster impact. It deals directly toward saving life, protecting property, search and rescue, providing emergency food, shelter and medical assistance. The effectiveness of responding to disaster is largely depends on the level of preparedness.

9.2 Role of Research Institutes, Organizations and Local Bodies in Disaster Risk Reduction

DRR is multifaceted and dynamic process that requires high potential and continuous efforts through scientific networks and technical initiatives, adjustments, decision-making process, and sharing and implementing the data of the existing disasters. For this purpose, interactions at different levels like key research institutes, government and non-government organizations, local bodies and communities should contribute for DRR to strengthen resilience for future disaster (World Bank 2001).

In India, Disaster Risk Reduction (DRR) initiatives has been played an important role—

(1) In institutional capacity building and risk mitigation investments at national, state, and local level.
(2) Community hazard and vulnerability mapping.
(3) Participatory planning and technology uptake.

Gujarat earthquake (January 26, 2001) and Orissa cyclone (1999) are good examples of establishing a proactive disaster risk management policy framework and risk mitigation efforts over the longer term. In follow-up, national cyclone risk mitigation projects covers all 13 cyclone prone costal and island/UTs. Stakeholders whose involvement is envisaged under this mission are India Meteorological Department, National Center for Medium Range Weather Forecasting, Ministry of Environmental and Forest, Central water Commission, National Institute of Disaster Management, Indian Institutes of Technologies.

The following are the key components—

(1) Upgradation of cyclone forecasting, tracking, and warning system.
(2) Computer processing capabilities.
(3) Cyclone risk mitigation and investments (construction of cyclone resistant shelters, shelter belt plantation, man-over regeneration.
(4) Technical assistance for Hazard Risk Management Capacity Building.
(5) Project management and monitoring.

9.3 Disaster Risk Reduction with Sustainable Future

There are many severe atmospheric mega disasters over the Himalayan region due to severe impact on the weather and climate conditions. It has unplanned urban growth and over exploitation of resources.

9.4 Himalayan Region

Have abundance of natural resources; contributing to hydrological cycle/geological activities.

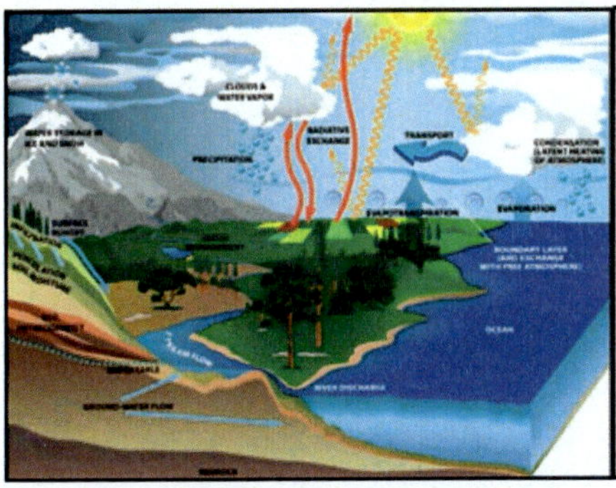

Hydrological Cycle

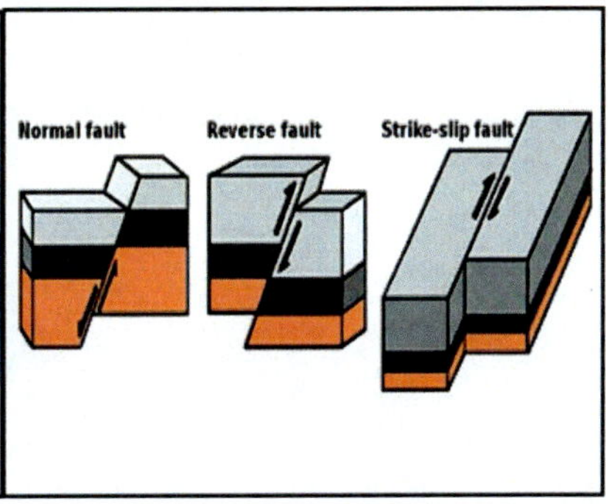

Plate tectonics

In June 2013, Uttarakhand flash flood associated with severe rainstorm and landslides caused more than 4000 casualties. The atmospheric warning will increase extreme rainstorms associated with landslides, flash floods and resulting huge number of glacier lakes, significantly increase the potential of natural disasters over this range. The economic and sociodevelopment of the South Asian countries are resulting the unplanned human intervention in nature, rising disaster risks and vulnerability in these areas. For DRR with sustainable future, it is necessary to the implementation of Early Warning System for the South Himalayan extreme weather.

Himalayas – Increase in natural disaster

Source: Uttarakhand flash flood – June 2013 report.

9.4.1 Sustainable Development—Issues Challenges and Opportunities

Rapid Urbanization and increasing population have multitude effect and cause threatening for increased energy use and environmental damage. The index for sustainable development is derived from the usage of natural resources and environmental degradation. Though it is cumbersome to discuss about all the factors affecting the sustainable development, an attempt for a brief description is made presented in this text.

The major threat to sustainable development is on account of growing population on the planet and the urbanization. Ecological footprint analysis is widely used around the world as an indicator of sustainable development. The kyoto protocol (2005) is to encourage the developed countries to reduce the carbon emissions and five other greenhouse gases or engage in emission trading if they maintain or increase emission of such gases. There may be a difference of opinion about the rate of climatic change but the fact remains that terms of area by 20% which is expected to cause a reduction of available water by 50% in terms of its volume. This may lead to the crises of food shortage and regional water conflicts, and there is a depletion of natural resources. In India, the shortage of water is felt on account of the effects of draughts and floods. The climatic change may shrink the Himalayan Glaciers in constrained economic development and environmental degradation. It is therefore imperative to exploit the available recycling options.

The smart/intelligent buildings, alternate power and fuel, and recycling options are so of the engineering solutions for sustainable development.

9.5 Contribution by IIT-Roorkee

Indian Institute of Technology Roorkee, Uttarakhand—India is the oldest technical institute ever since the period of British rules. The Institute started its journey with the construction of famous Ganga Canal (1840). Since then the Institute has made outstanding contribution in the major areas of civil engineering for national development (e.g., large/small dams, irrigation engineering, earthquake engineering, infrastructure engineering, wind engineering, water resources, etc.).

Having attained significant strength of civil engineering and to meet the present need of strengthening, the safety issues against the natural and man-made hazards, the institute established a multidisciplinary "Centre of Excellence in Disaster Mitigation and Management" in the year 2006 and presently it runs a full time post graduate program (M.Tech) and doctoral program (Ph.D.) since 2012.

The main aim of conduction multidisciplinary research and educational program is to create a national database for rapid dissemination of information and knowledge on technical and managerial skills to professionals to make them equipped with innovative technologies for risk mitigation and management of disasters for overall benefit of the society. The center aims to achieve excellence in the key areas especially geological induced risk studies and wind induced risk studies and their early warnings.

9.6 Summary

The efforts by Disaster Prevention Research Institute Kyoto University has played a vital role in bringing researchers and research institutes from all around the globe on a platform to develop a fruitful alliances in an effort of Disaster Risk Reduction.

The Centre of Excellence in Disaster Mitigation and Management at Indian Institute of Technology is expected to take a lead role in this direction.

References

Alexander D (2000) Confronting catastrophe: new perspectives on natural disasters. Oxford University Press, New York

World Bank (2002) The last straw: integrating natural disaster mitigation with environmental management, prepared by van Aalst M, Burton I, Disaster management facility, working paper series No. 5

World Bank (2001) Social protection sector strategy: from safety net to springboard. The World Bank, Washington, D.C.

Chapter 10
Evaluating the Success of Participatory Flood Risk Mapping—A Case Study from Dharavi, Mumbai

Subhajyoti Samaddar, Hirokazu Tatano, and Ram Sateesh Pasupuleti

Abstract In recent decades, several participatory tools and techniques have been introduced in disaster risk management. These tools vary from each other at great extent in terms of their structure, function, scope, facilitation process, required time and resources and so on; however, the sole objective of all of them is to ensure effective involvement of local communities in disaster risk management. If involving community in the decision-making process is simple and unidimensional, then one particular tool would be enough to ensure the objective. However, the multifaceted nature of community participation, the variety of ways to operationalize them and its numerous untested claims have created enormous confusion amongst researchers and practitioners to put the idea into practice. There is urgent need to systematically define the objectives of the community-based disaster risk management and examine how and what extent the participatory tools can ascertain those objectives. To address these issues, this study introduces a participatory flood risk mapping exercise in Mumbai, India. In this study, we unraveled how and what extent the participatory flood risk mapping exercise is effective to achieve the critical parameters of good participation. Thereafter, local community themselves evaluated the success of the participatory risk mapping to ascertain various participation yardsticks. This study is one of the pioneering works in identifying the factors accountable for effective public participation in disaster management and to validate them empirically.

Keywords Participatory risk mapping · Evaluation · Flood · Mumbai

10.1 Introduction

Participatory risk mapping is considered as a powerful tool to engage local communities in disaster risk management. Participatory risk mapping entails a process that

S. Samaddar (✉) · H. Tatano
Disaster Prevention Research Institute, Kyoto University, Kyoto, Japan
e-mail: samaddar@imdr.dpri.kyoto-u.ac.jp

R. S. Pasupuleti
Department of Architecture & Planning, Indian Institute of Technology Roorkee, Roorkee, India

H. Tatano and A. Collins (eds.), *Proceedings of the 3rd Global Summit of Research Institutes for Disaster Risk Reduction*, Disaster and Risk Research: GADRI Book Series,
https://doi.org/10.1007/978-981-15-8662-0_10

enables local community themselves, in collaboration with experts or researchers, to physically locate—their disaster risks, the resources they possess and the actions they can take to reduce vulnerabilities (Osti et al. 2008; White et al. 2010). As the promotion of disaster resilient society requires a paradigm shift away from the primary focus on natural hazards and their quantification toward identification, assessment and ranking of various vulnerabilities, the identification of spatial risks from local communities' perspectives is critical and prerequisite (Gaillard and Pangilinan 2010). Spatial features or conditions of the settlement are the expression of dynamic interaction between natural and cultural forces in the environment (Pasupuleti 2013). Cultural landscapes are the result of consecutive reorganization of the land in order to adapt its use and spatial structure better to changing social demands (Pasupuleti 2012). Spatial mapping allows cross-validation from other maps like contour, land use, etc. Local community's deep understanding about the locality has considerable value, as well as supplementary and cross-validating knowledge, for understanding disaster situation and designing community-based amelioration (Tran et al. 2009; Cadag and Gaillard 2012). Affected communities often have a comparative advantage in understanding the etiology, consequences and nuance of complex problem (Haynes et al. 2007; Reichel and Frömming 2014). This is because, undoubtedly, it is the local community that is both the primary victim and the first to respond to emergencies when disaster strikes (Tran et al. 2009; Samaddar et al. 2011a, b). So, involving local community is a perquisite for effective implementation of disaster risk reduction strategies. Participatory risk mapping fosters the participation of threatened community in both the evaluation of risk and the ways to reduce it (Chingombe et al. 2015). Also, sharing direct experiences from affected community helps to pinpoint the problems at the grassroots level and hence ensures successful implementation of strategies (Newig et al. 2016). Participatory risk mapping plays a critical role in risk awareness because it enables visual representation of risk faced by the communities and the strategies they propose for disaster risk reduction (Gaillard and Maceda 2009). Participatory risk mapping is now also being promoted by local governments in order to strengthen the links between the official disaster management plans and community-based disaster management (Newig et al. 2016). When community participation is ensured in the process of spatial risk mapping, it enhances the acceptability of disaster risk reduction strategies. The exercise presumably escalates the active involvement of stakeholders through allowing them to express their ideas and perspectives in an easily understandable visual format (Reichel and Frömming 2014; Liu et al. 2018).

The significance of participatory risk mapping has been widely recognized; however, it remains conceptually uncertain and undefined how this tool can ensure the meaningful engagement of local communities in disaster risk management (Osti et al. 2008). There is no universal yardstick for conducting participatory risk mapping, but different objectives, steps, facilitation process, art of conducts have been adopted by different organizations across regions (e.g., see Samaddar et al. 2011a; b; Cadag and Gaillard 2012; Reichel and Frömming 2014; Chingombe et al. 2015; Liu et al. 2018). Moreover, oftentimes, one claims better results than others. However, by principle, they all remain one, doing participatory disaster risk mapping. Being untested, this powerful tool raises doubts about its effectiveness (Haynes et al. 2007). Voluminous studies have been done on disaster risk mapping; nevertheless, there is no sufficient amount of scientific literature that suggested a comprehensive framework for participatory risk mapping. Conceivably, there lies a presumption amongst researchers that the participatory exercise is just a means to an end; therefore, introducing it would automatically involve the community in the decision-making process. In the absence of defined process framework, the exercise leaves practitioners remain clueless about how, when and what extent a local community can be meaningfully involved in disaster risk management? What are the lacunas found in previous exercises and how they can be improved? Till date, barely any study systematically delineates the process objectives of participatory risk mapping and the mechanism to realize them. Consequently, it has been observed that successful community involvement through participatory risk mapping remains localized and has never been successfully scaled up or replicated in other place (Samodra et al. 2018). The effectiveness of participatory risk mapping appears to be self-proclaimed. After spending lots of money and time, the vulnerable and marginalized sections of the community remain unsolicited and unheard (Samaddar et al. 2019). This conundrum has the tendency of affecting the enthusiasm of practitioners and planners in adopting participatory disaster management tool (Samaddar et al. 2015b). There is a need to objectively define the structure and process of the participatory risk mapping and to leash out how and what extent the adhered processes would be effective to accomplish the objectives.

To address these issues, this study will introduce a participatory flood risk mapping exercise in Mumbai, India, to systematically define how and what extent this tool can potentially meet the goals of effective community-based disaster management. This study is divided into two parts. First, we introduce a participatory risk mapping exercise carried in a flood-prone informal settlement in Dharavi, Mumbai. We elucidate how we designed and implemented the participatory flood risk mapping in order to adhere the process criteria of effective participation. In the second part, this study shows how the local communities evaluate the success and effectiveness of the flood risk mapping in terms of their participation. This study in general will help us to identify the critical processes of community participation in risk mapping.

10.2 Process Mechanism of Community Involvement in Participatory Risk Mapping

An effective participatory mechanism has to ascertain a process that ensures the successful community involvement in the decision-making process (Samaddar et al. 2015a; b). There exists no single definition of participation, and the concept remains contested (Chess and Purcell 1999). Participation has been understood variedly and given different names (Arnstein 1969; Rowe and Frewer 2000; Samaddar and Okada 2006). Similarly, carrying out participatory exercises is considered as an art, where the communication skill of the facilitator, the place and timing of the exercise, language used, all matter to decide how effectively the exercise can involve participants in the process (Cronin et al. 2004; Na et al. 2009; Okada et al. 2013). Defining effective process for community-based disaster management could be value laden; therefore, a rigorous and careful definition is needed. The process means the quality and characteristics of the means of participation. The participation process decides what facilitation process to adopt, whom to involve, how, when and what extent and so on (Rowe and Frewer 2005; Chess and Purcell 1999; Webler et al. 1995). Although local communities' participation has been widely acknowledged in disaster risk management, just how to involve them remains controversial (Renn et al. 1993). Therefore, the criteria of participation process should be systematically defined in order to delineate the steps, facilitation process and timing of community involvement. There exist only a few frameworks for community participation in disaster risk management (Bajek et al. 2008; Samaddar et al. 2015a, b). In this study, we used the participatory framework for disaster risk management proposed by Samaddar et al. (2018) to define the process criteria of community participation in risk mapping exercise as listed out in Table 10.1.

10.3 Participatory Flood Risk Mapping in Kalaqila, Dharavi, Mumbai

The participatory flood risk mapping exercise was carried in a small flood-prone informal settlement, known as Kalaqila Chawl, in Dharavi area of Mumbai city, India (see Fig. 10.1). The objective of the exercise was to identify the spatial flood risks of the area as perceived by the local community. It was organized under the GCOE-HSE Program (Global Center of Excellence Program on – Human Security Engineering for Asian Megacities) carried out by the Kyoto University, Japan, in collaboration with local city government, Municipal Corporation of Greater Mumbai (MCGM) and other academic institutes including School of Planning and Architecture (SPA), New Delhi, and Sir J. J. College of Architecture, Mumbai.

Mumbai, the financial capital of India, is considered as one of the most vulnerable coastal megacities in the world. The increasing vulnerability of Mumbai city is attributed with huge land reclamation, vegetation clearances and abrupt land use changes which took place in order to adjust with its rapid urbanization, industrialization and population growth over the decades (Bhagat et al. 2006; Yadav and Desai, Year Not Mentioned; Samaddar et al. 2011b). On July, 2005, the city experienced catastrophic flood due to heavy rainfall and high tidal waves (Gupta 2007; Samaddar et al. 2014). The flood killed around 900 people and damaged more than 90,000 houses (Bohra et al. 2006; Government of Maharashtra 2006; Samaddar et al. 2012a, b). The total losses rose to around \$3–5 billions (Tatano and Samaddar 2010). The poor, marginalized communities living in squatter settlements were the most severely affected populations of Mumbai (Parthasarathy 2009).

Table 10.1 Process criteria of participation and the account of local communities' involvement in flood risk mapping in Kalaqila community, Mumbai

Process criteria of participation	Local community's involvement in participatory flood risk mapping, Kalaqila, Mumbai
Early engagement: Local communities should be involved from the very beginning of the project and identify the issue	It was revealed from the pilot survey that the Chawl or neighborhood leaders play critical role for community mobilization and collective decision-making process in these communities. The research team first communicated all Chawl leaders and informal about the tentative purpose of the project. Over the period of time, the rapport and trust building with local leaders had been initiated. For this, the research team made weekly visits to the community and organized several informal group meetings. The research team gradually proposed the aim and objectives of the flood risk mapping to local leaders
Representation of relevant stakeholders: All groups and sections of the community should actively participate	We considered Chawl leaders as the true representatives of the community and engaged them throughout the mapping exercises, like area selection, site appraisal and field surveys and preparing maps
Clear and agreed objectives: The project should have a clear objective and the community should agree to work on it	The local community had been informed that the main purpose of the exercise was to collect flood risk information of area to enhance local community's risk awareness. The actual implementation of the project started only after the local community agreed to proceed further
Continued and active engagement: Local communities should be actively engaged in all stages of the decision-making process	Community leaders helped researchers in identifying the area, establishing links with community members and revise exiting base maps. A large number of community members actively participated in flood mapping through identifying the vulnerable areas, flood height and duration in different places through town watching, observations, group discussions and open-ended interviews. Local leaders also provided critical inputs in revising flood risk maps
Fairness: Local communities should enjoy the freedom to propose any topic for discussion and express their opinions freely	There was no reservation for active participation of local community. The risk mapping exercise including town watching, group discussions and interviews was carried out inside the community. Participants enjoyed the freedom to express their observations, opinions and concerns

(continued)

Table 10.1 (continued)

Process criteria of participation	Local community's involvement in participatory flood risk mapping, Kalaqila, Mumbai
Power to influence decisions: Local communities' suggestions and observations should be reflected in the project plan. Community should enjoy the power to influence the project decision	Local communities were given the highest priority in the meetings, group discussion and survey to locate the flood vulnerable areas and buildings, flood damage and loss, magnitude of flood, waterlogging areas and potential countermeasures. Local leaders played critical role to design the fieldwork plan
Capacity building: Local communities should be educated and trained for knowledge and skill upgradation. The capacity building initiative should enable local communities to meaningfully involve in debates and discussions with other stakeholders in the decision-making process	Students with architecture and planning background helped local leaders how to read and draw maps. It helped local people to record the flood information on maps
Incorporating local knowledge: Indigenous knowledge, local resources and expertise should be used in the project	The flood risk maps were developed solely based on the information provided by local communities from their own experience and day-to-day observations
Good facilitation: Often face-to-face open discussion in local language considered to be most effective ways to interact with the community. Good facilitators with lots of experience should be involved to carry out the participatory exercise	We used town watching, group discussion, face-to-face dialogues, oral history to collect information for preparing flood risk maps. All exercises were conducted in Hindi which is the most popular language in this community Students from architecture and planning discipline (from School of Planning and Architecture, New Delhi, and Sir J. J. College of Architecture, Mumbai) helped local communities in field surveys and recoding information on maps
Resource availability: There should be adequate natural and financial resources to carry out community-led initiatives	The costs for organizing surveys, map digitizing and printing were covered by the research group. Community provided their local community halls and furniture to hold meetings and group discussions. No one was paid for their participation or value consultation
Time: There should be a time frame for project implantation	The exercise was carried out on Sunday and other holidays when working populations of the community can participate. It took two months to finish the risk mapping

Note The process criteria of participation were derived from the participatory framework proposed by Samaddar et al. (2018)

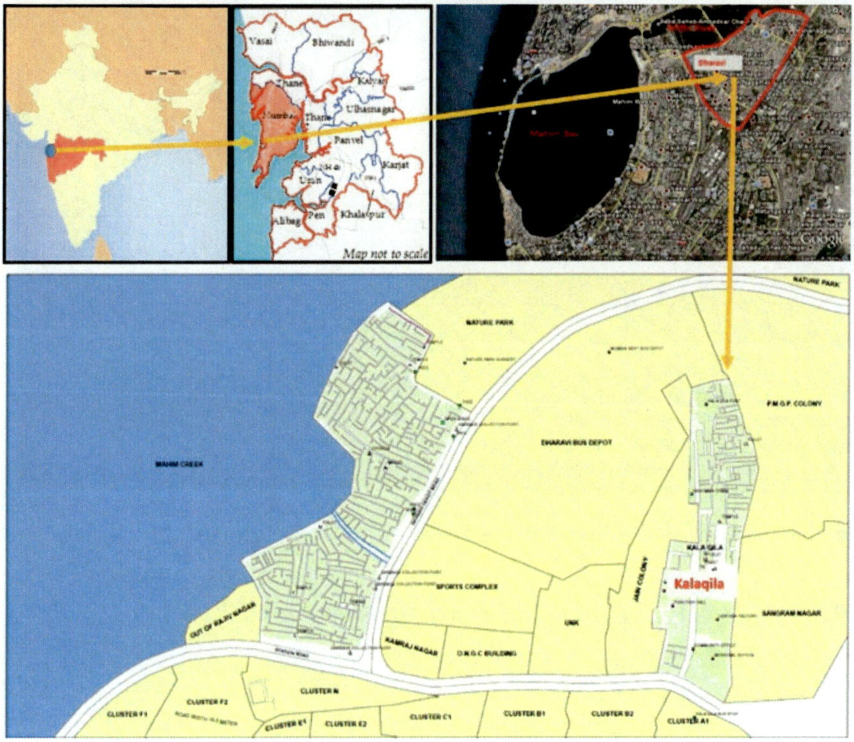

Fig. 10.1 Location map of Kalaqila Chawl, Dharavi

The study area, Kalaqila Chawl (see Fig. 10.2), is located in Dharavi which is infamously known as the largest slum in Asia. Back in the 1930s, Dharavi was a small fisherman village located at city's outskirt on the bank of Mithi River. In the 1960s and 70s, during the boom of cotton industries, millions of landless rural people migrated to Mumbai in search of jobs. The formal housing options were unaffordable for this newly migrated labor communities. The hazardous, environmentally fragile and abandoned settlements like Dharavi became the prime hot-spots for these marginalized and poor populations to settle down in Mumbai (Parthasarathy 2009; Samaddar et al. 2012b). Over the decades, more and more wetlands had been filled up and reclaimed by the poor migrates, and extremely dense and complex living quarters gradually developed in Dharavi. Today, more than half a million people live in just 1.75 sq.km in Dharavi. Kalaqila Chawl is one of the oldest settlements in Dharavi. It houses 700 inhabitants. The local municipal office (G-North Ward, Municipal Corporation of Greater Mumbai) features the area as a low-lying, chronic flood-prone area. The annual waterlogging and small-scale flooding due to heavy rain and high tight combination are common annual phenomena throughout the rainy seasons in those areas.

Fig. 10.2 A ground view of Kalaqila Chawl, Dharavi

10.3.1 Steps and Process of Participatory Flood Risk Mapping

The steps followed to carry out the flood risk mapping exercise in Kalaqila Chawl, Dharavi, are given in Table 10.2.

10.3.2 Tools and Techniques of Flood Risk Mapping

A large range of tools were used for data collection and mapping. Some of the key tools used in this exercise are as follows.

Town watching: Citizens joined the research team to walk through the settlement and record the flood risk information on maps. Town watching helped to record the flood history and its impact, to identify the linkage between housing characteristics, settlement patterns and infrastructure facilities with the increasing flood risk in the community (see Fig. 10.3).

Observations: Observation surveys were carried out to collect information in regard to build environment, land use and quality of community infrastructures including water supply, electricity, sewerage, road and so on (see Fig. 10.4).

Open-ended interviews were used for collecting data on socioeconomic impacts of flood, property loss, duration and height of flood water in settlement clusters. It also helped the key informants to elaborate their concerns and flood experience. In each lane, two face-to-face open-ended interviews were conducted.

Table 10.2 Steps and processes of participatory flood risk mapping in Kalaqila Chawl, Dharavi

Steps	Brief description
Step-I: problem identification and goal setting	The researchers (under the GCOE-HSE Program) decided to carry out participatory mapping exercises to identify flood risks in Kalaqila Chawl to enhance community's risk awareness
Step-II: site selection	Based on the suggestion from Municipal Corporation of Greater Mumbai (MCGM), Kalaqila Chawl, one of the most flood-prone settlements in Dharavi, was selected as case study area
Step-III: identifying stakeholders and local leaders	Field engineers of MCGM helped the research team to identify local leaders of Kalaqila and develop initial rapport with them. Afterward, local leaders helped researchers to get acquainted with local residents, different religious and ethnic (linguistic) representatives in the community
Step-IV: rapport building	The next step was to build relationship with local inhabitants. Kyoto University GCOE-HSE research team regularly visited local leaders and conducted informal group meetings to exchange information and concerns about waterlogging and inundation issues in the community
Step-V: plan proposal for participatory flood risk mapping	The research team explained the aim and objectives of the projects to local leaders and solicited their active participation. Local leaders agreed to collaborate to conduct the flood risk mapping exercise
Step-VI: base map preparation	Initially, there was no layout map of the area. After sometime, a base map was received from MASHAL NGO that conducted a spatial survey of the area for the project, called Dharavi Redevelopment Plan. The base map was then revised and updated based on the information received from local leaders
Step-VII: orientation and training of the local leaders for mapping exercise	While revising the baseman, it was realized that the local leaders could not read maps. Then, with the help of architecture and planning students (from Sir J. J. College of Architecture), a short workshop was conducted to help local communities to read and record information on maps

(continued)

Table 10.2 (continued)

Steps	Brief description
Step-VIII: finalizing time and date of the exercise and communication with the community	After the base map preparation, the dates and time of the exercise were decided. Keeping in mind the prospect of participation of all interested citizens, it was decided that the exercise would be conducted on only Sundays and other holidays. The Chawl (neighborhood) leaders took the initiative to communicate with their fellow citizens for their participation
Step-IX: flood risk survey and mapping	The flood risk mapping exercise was carried out through the collaborative efforts made by citizens, local leaders, researchers and students, and field engineers. For the flood mapping, a wide range of survey tools including town watching, group discussion, household interviews, observations were carried out
Step-X: map digitization	Once the data recoded on maps, they were digitized by students and research team
Step-XI: data validation by local leaders	Digitized maps were given to the local leaders for correction and filling up missing information
Step-XII: display the map to the local leaders and revising the map	Flood risk maps were displayed in Chawls (neighborhoods) for taking comments and observations of all community members for further revision
Step-XIII: completion of flood risk maps	Final risk maps had been prepared covering a wide range of flood risk aspects—flood heights, flood duration, vulnerable areas, damage and losses of 2005 flood, waterlogging areas and community's adaptation strategies
Step-XIV: information circulation	Printed copies of flood risk maps were distributed to the Chawl leaders to circulate and display it in Chawls and community

Group Discussion: For each Chawl (neighborhood), two group discussions were organized. Group discussions were carried out through informal group meetings with local communities to exchange ideas, observations and experience about local flood issues. Group discussions helped to identify flood-affected areas, property damage as well as coping strategies of residents (see Fig. 10.5).

Photography: Photographs played a critical role through gathering evidence on flood risk in the community. Photographs helped at a later stage to identify a location and its physical conditions and aspects.

Secondary Data Collection: Maps or data were also collected from secondary sources such as government records (e.g., Municipal Corporation of Greater Mumbai (MCGM)) and NGOs (e.g., MASHAL).

Fig. 10.3 A moment of town watching in Kalaqila Chawl

Fig. 10.4 During the observation session in Kalaqila Chawl

Fig. 10.5 Local communities discussing with researcher the flood risks in their community during a group discussion session in Kalaqila Chawl, Dharavi

Mapping: Mapping generally includes physical aspects rather than social or economic aspects. But in the present exercise, zones of socioeconomic vulnerability were also be physically represented and demarcated on map. All maps were digitized by the research team.

10.3.3 Data Collection and Mapping

The flood risk mapping exercise collected the following sets of data—

(i) Land use—commercial, residential, public–semipublic, playground/parks, water bodies and land cover.
(ii) Infrastructures—water supply networks and facilities, electricity lines, road networks and condition, community toilet, drains/nullah, waste dumping areas, religious buildings, doctor clinics, school, etc.
(iii) Flood—frequency, duration, height of past flood events, waterlogging sites, property damage and household loss, perceived vulnerable areas and groups.

10.3.4 Roles and Involvement of Stakeholders

The flood risk mapping exercise was mainly community-led, but other stakeholders had also actively participated. Table 10.3 depicts the involvement of different stakeholders in different phases of the exercise. Table 10.1 shows the process mechanism that was adopted in this exercise to ensure the meaningful involvement of the local communities. The process-based criteria of community participation is adopted from the participatory framework proposed by Samaddar et al. (2018).

Table 10.3 Roles and involvement of the stakeholders including local communities in the flood risk mapping at Kalaqila Chawl, Dharavi

Stakeholders/participants	Number of participant and representation	Time of involvement	Roles and nature of involvement
Local community	9 local leaders from different Chawls (neighborhoods) Participation of common citizens in different phrases of the exercise	Local leaders—from problem identification and project goal settings Citizens—from risk mapping and field survey	• Plan preparation • Revising base maps • Main source of information for flood mapping • Helping to finalize flood maps
Research team	8 members from GCOE-HSE research team including Kyoto University (Japan), School of Planning and Architecture (New Delhi), Sir J. J. College of Architecture (Mumbai)	Proposal and site identification	• Goal settings • Site appraisal • Stakeholder identification and rapport building • Facilitating field surveys, town watching and map preparation
Students	Total 15–20 students' participation from—School of Planning and Architecture, New Delhi, and Sir J. J. College of Architecture, Mumbai	Mapping	• Revising base maps • Giving training of local leaders for reading and recording information on maps • Facilitating group discussion and town watching • Map digitization
Local government	Field engineers and official of G-North Ward, Municipal Corporation of Greater Mumbai (MCGM)	Town watching	• Identifying local leaders • Site selection • Providing critical information

10.3.5 Fact-Findings from Flood Risk Mapping

Kalaqila is predominately a residential area (Fig. 10.6). The civic amenities and development works are skeletal and totally unplanned manner. Poor drainage capacity, inadequate solid waste management and unsafe electricity connections all are vehicles of transporting risks during and after the flood to the area (see Fig. 10.7). The capacity of the garbage bins is limited; therefore, the bin gets overfilled and the garbage gets spread around. People are careless, and they throw the garbage in the gutter. Due to this, the gutter gets choked and causes problem in water flow; then, there comes nasty smell from the gutter. All these activities cause health and environmental problems. Drinking water supply pipes and faucets are generally set at short height from the ground, and so these facilities got inundated by flood or waterlogging during the monsoon. People do not get potable water throughout the monsoon season, especially during waterlogging phases. Lanes within the settlement are narrow, rugged, zigzag and unpaved. It is even difficult to move in ordinary days (see Fig. 10.7).

In rainy season when there are heavy rains and high tides, the drainage gets filled very soon and starts overflowing. This water enters the houses and causes damage to household goods. The community experience several waterlogging during monsoon from June to August (see Fig. 10.8). The MCGM is irregular in collecting the garbage. During waterlogging when the lanes are inundated, the areas become a death trap. Cables are open and very old. Inhabitants fear that any time a short circuit can happen and kill people. Like any other parts of Dharavi, in Kalaqila also, there are many industries and businesses which create waste and pollution (see Figs. 10.6 and 10.7). Gutter is kept open for the water to flow but that bring dirt and wastes. The 2005 flood was unprecedented. Some parts of the area experienced flood height of 3–4 feet. A vast area of the slum had flood water for 48 h (see Fig. 10.9). The loss due to the 2005 flood was enormous. Figure 10.9 shows that more than half of the population had 10,000–20,000 INR household loss due to the flood. Prior to 2005, people did not have any experience of the flood of that magnitude. Most citizen in the community did not foresee the magnitude of the flood and, therefore, reluctant to evacuate. There was no evacuation announcement, but some resident instantaneously moved to nearby higher places by their own. Evacuation was the key concern for a significant portion of the participants. The absence of early warning system, reliable information and designated evacuation shelter had been considered main barriers for timely evacuation. There are many tall public and private buildings around the area, which can be used for evacuation shelter if access to those shelters can be ensured. Participants believe their own social networks including religious groups and local clubs as well as the friendly and helpful mentality of local people are lifeline for slum dwellers to survive during flood emergency.

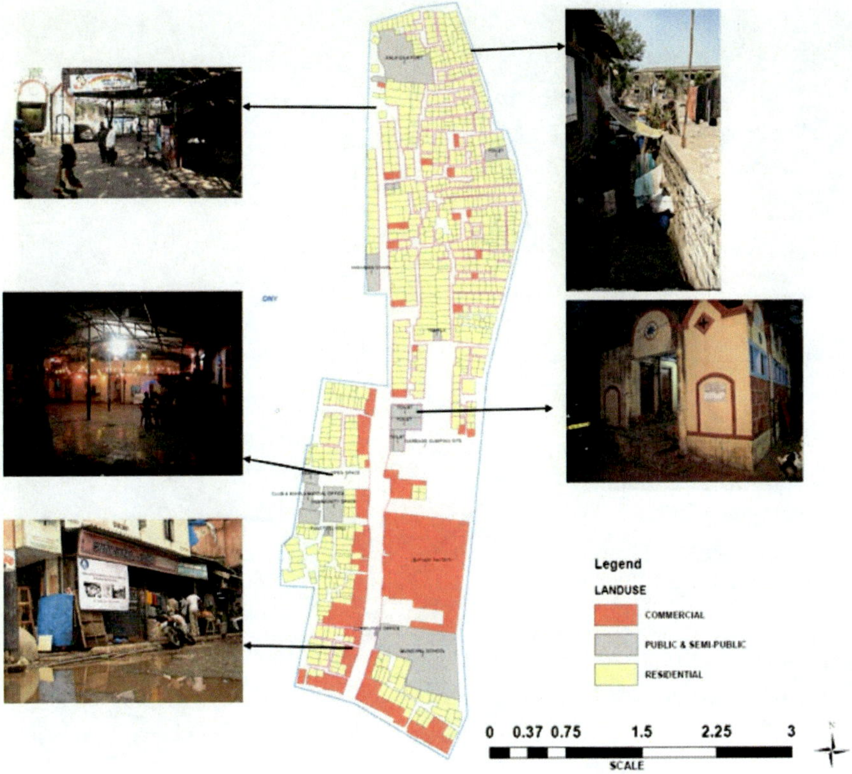

Fig. 10.6 Land use distribution at Kalaqila Chawl

10.4 Evaluation of the Effectiveness of Participatory Flood Risk Mapping

In order to evaluate the success of the participatory flood risk mapping from the local community's perspective, after the exercise was over, we interviewed 19 local community members who participated in the exercise. The participants were asked to evaluate the achievement of the exercise in terms of community participation. The scores were received in 10 points scale where 1 is "no achievement" and 10 "fully achieved." The interviews were face-to-face and taken in Hindi language few weeks after the exercise.

Figure 10.10 shows evaluation score of the flood risk mapping for effective community involvement. Participants reported that five process-based criteria of participation had been well achieved by this exercise—early engagement, clear objectives, fairness, power to influence decisions and incorporating local knowledge and understanding. But, three criteria of participation including representation of relevant stakeholders, capacity building and resource viability had been minimally achieved.

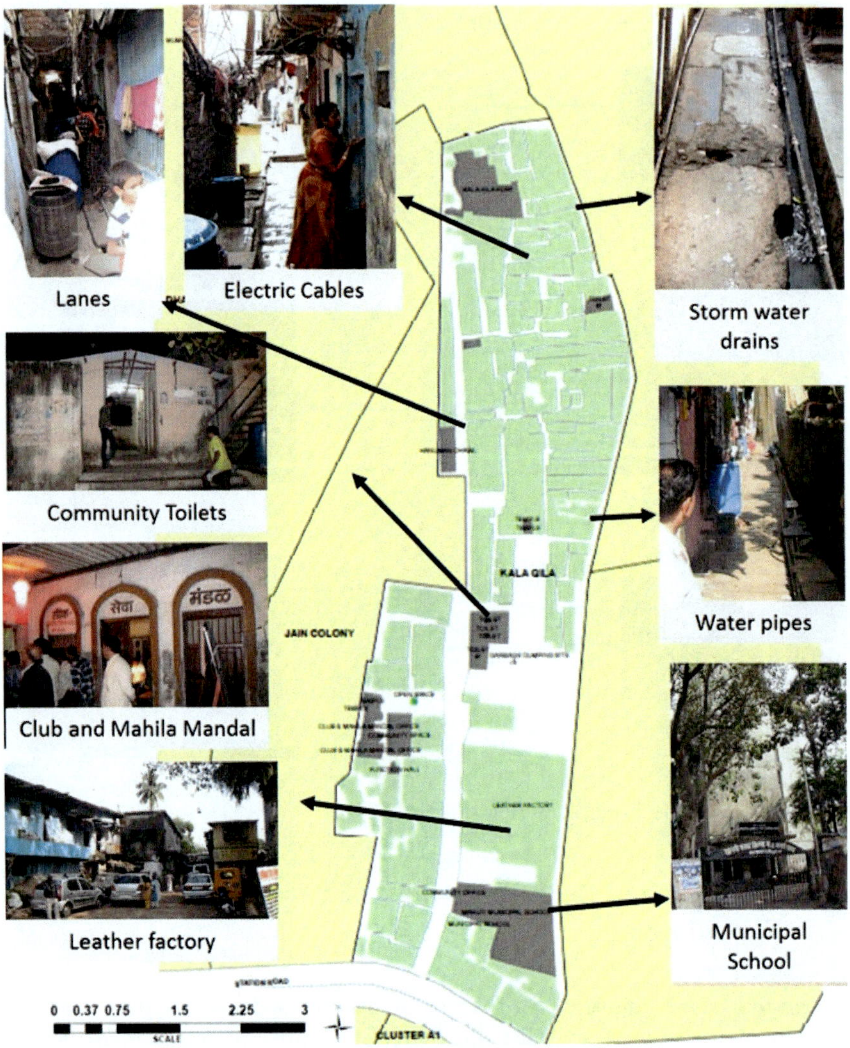

Fig. 10.7 Social and physical infrastructure at Kalaqila Chawl

Items received moderate score are continued and active engagement of the community and good facilitation. Community believes that the research team was successful to make the objectives of the exercise very clear to the community as well as the plan was well executed. Moreover, community reported that they enjoyed the freedom of participation and the power to influence the decisions of the project. Local communities, however, believe that all sections of the community did not show interest in the exercise. The leaders who actively involved in the exercise do not truly represent the total population of the community. The project information should have diffused

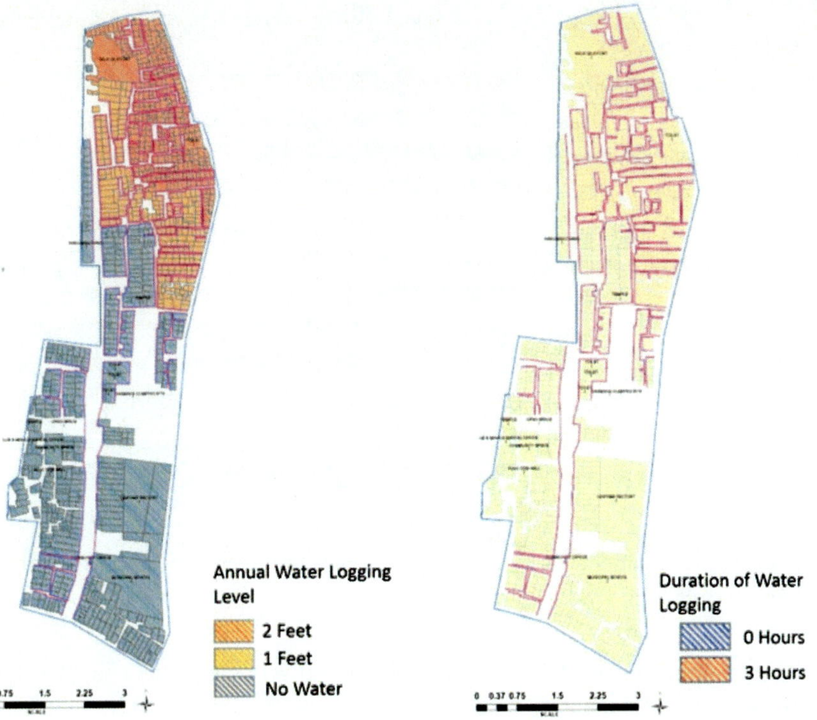

Fig. 10.8 Annual waterlogging level and duration in Kalaqila Chawl, Dharavi

Fig. 10.9 Magnitude and impact of 2005 flood in Kalaqila Chawl, Dharavi

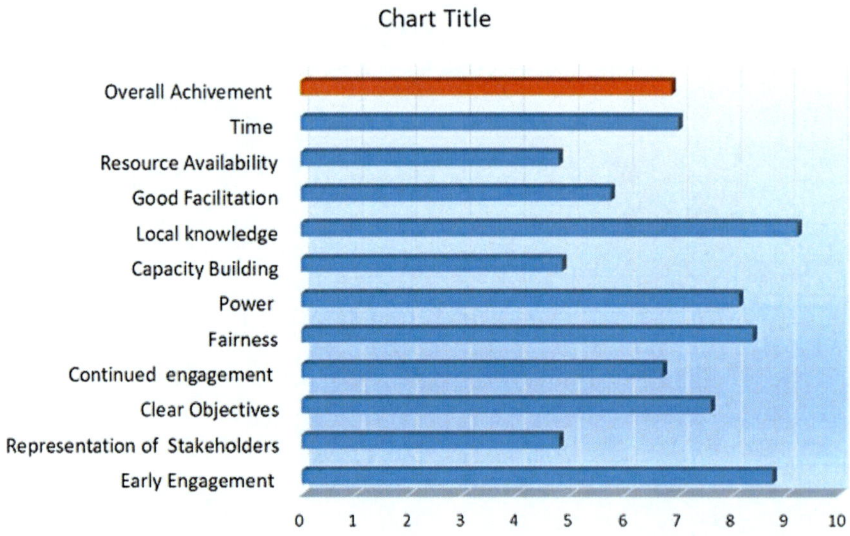

Fig. 10.10 Community's evaluation of the effectiveness of participatory flood risk making for meaningful community involvement

to larger communities, and the participation of all sections of the community was expected. Some community members initially showed interest in the project and attended few meetings, but afterward discontinued their participation. Therefore, it can be concluded that the continued engagement of the community throughout the exercise was moderately achieved. Participants reported that the exercise did not enhance the true capacity of the community because the local community was still not self-dependent. Prevailing illiteracy, lack of technical skill and knowledge in addition to poor finical resources impede the local community to carry out any community-led plans. Without the help of external agencies, the local community would not be able to organize similar exercise again. Participants were satisfied with the facilitation process of flood risk mapping. But they found that it is extremely challenging to record flood risk information on maps. Overall performance of the exercise was marked as high. Participants believed that the flood risk mapping exercise was able to achieve the major objectives of the exercise.

10.5 Conclusions

Community participation becomes a commonplace element in disaster risk management, yet there exists no consensus how to successfully involve the local communities in the decision-making process (Samaddar et al. 2019). In recent decades, therefore, several participatory tools and techniques have been introduced in disaster risk management to ensure community participation in the decision-making process.

Each of them appears promising and claims better result (Na et al. 2009). But the planners and practitioners in the field are confused and challenged to decide which participatory tool to select and what basis (Cronin et al. 2004)? Moreover, in reality, the effective participation of local communities remains elusive. The selection of appropriate tools and techniques is a daunting task because community participation is understood and practiced variedly.

For long, the participatory disaster risk mapping has been advocated as an effective tool for local community's involvement in disaster risk reduction. Today, it is considered as the most popular and widely employed participatory disaster management tool (Osti et al. 2008; Gaillard and Pangilinan 2010). Voluminous studies have done on participatory disaster risk mapping, but none of them have ever suggested a comprehensive framework for participation. In the absence of defined process framework, the exercise leaves practitioners remain clueless about how, when and what extent a local community can be meaningfully involved in disaster risk management. What are the lacunas found in previous exercises and how they can be improved? Conceivably, there lies a presumption amongst researchers that the participatory exercise is just a means to an end; therefore, introducing it would automatically involve the community in the decision-making process. According to authors' knowledge, till date, barely any study systematically has delineated the process objectives of participatory risk mapping and proposed the mechanism to realize them. Consequently, it has been observed that successful community involvement through participatory risk mapping remains localized and has never been successfully scaled up or replicated in other place (Samodra 2018). The effectiveness of participatory risk mapping appears to be self-proclaimed. After speeding lots of money and time, the vulnerable and marginalized sections of the community remain unsolicited and unheard (Samaddar et al. 2015b). This conundrum has the tendency of affecting the enthusiasm of practitioners and planners in adopting participatory disaster management tool (Samaddar and Okada 2006). There is a need to objectively define the structure and process of the participatory risk mapping and to leash out how and what extent the adhered processes would be effective to accomplish the objectives.

Addressing this research question, we introduced a participatory flood risk mapping exercise in Mumbai, India, to systematically define how and what extent this tool can potentially meet the objectives of successful community-based disaster management. Based on a proposed participatory framework by Samaddar et al. 2018, we first identified the process criteria of participation. As the process objectives were set, the structure and function of the exercise were more rigorously defined to transfer the concept into practice. The exercise was meant to enable the effective participation of local communities. Therefore, in order to understand its effectiveness, this participatory exercise was evaluated by the end-users or participants. The participants rated the effectiveness of the tool in order to achieve the desired process-based criteria of community participation. To evaluate the effectiveness of the participatory spatial risk mapping, after the exercise, we interviewed the local community leaders who participated throughout this exercise. The results revealed that the local community valued very high achievement of five process criteria of participation through the exercise as follows: early engagement, clear objectives, fairness, power to influence decisions, incorporating local knowledge and understanding. According to the respondents, the participation criteria including representation of relevant stakeholders, capacity building and resource viability had been minimally addressed in the flood risk mapping. The exercise had the moderate-level success to achieve the following process objectives of community participation—continued and active engagement of the community and good facilitation. So, community believed that major process objectives of the community-based disaster risk management were achieved by the participatory flood risk management in Kalaqila Chawl. There exists rare, perhaps no literature in the disaster risk management studies, which systematically define how participatory tools would enable the meaningful involvement of the community in the decision-making process. This study is one of the pioneering works in identifying the factors accountable for effective public participation in disaster management and validating them empirically.

References

Arnstein SR (1969) A ladder of citizen participation. J Am Inst Plan 35(4):216–224

Bajek R, Matsuda Y, Okada N (2008) Japan's Jishu-bosai-soshiki community activities: analysis of its role in participatory community disaster risk management. Nat Hazards 44(2):281–292

Bhagat RB, Guha M, Chattopadhyay A (2006) Mumbai after 26/7 deluge: issues and concerns in urban planning. https://doi.org/10.1007/s11111-006-0028-z. Retrieved from https://www.springerlink.com

Bohra AK, Basu S, Rajagopal EN, Iyengar GR, Das-Gupta M, Ashrit R, Athiyaman B (2006) Heavy rainfall episode over Mumbai on 26 July 2005: assessment of NWP guidance. Current Sci 90(9):1188–1194

Cadag JRD, Gaillard JC (2012) Integrating knowledge and actions in disaster risk reduction: the contribution of participatory mapping. Area 44(1):100–109

Chess C, Purcell K (1999) Public participation and the environment: do we know what works? Environ Sci Technol 33(16):2685–2692

Chingombe W, Pedzisai E, Manatsa D, Mukwada G, Taru P (2015) A participatory approach in GIS data collection for flood risk management, Muzarabani district Zimbabwe. Arab J Geosci 8(2):1029–1040

Cronin SJ, Petterson MG, Taylor PW, Biliki R (2004) Maximising multi-stakeholder participation in government and community volcanic hazard management programs; a case study from Savo Solomon islands. Nat Hazards 33(1):105–136

Gaillard JC, Pangilinan MLCJD (2010) Participatory mapping for raising disaster risk awareness among the youth. J Contingencies Crisis Manage 18(3):175–179

Gaillard JC, Maceda EA (2009) Participatory three-dimensional mapping for disaster risk reduction. Participatory Learn Action 60(1):109–118

Government of Maharashtra (2006) Report of the Fact Finding Committee (FFC) on Mumbai Floods

Gupta K (2007) Urban flood resiliency planning and management and lessons for the future: a case study of Mumbai. Urban Water J 4(27):183–194

Haynes K, Barclay J, Pidgeon N (2007) Volcanic hazard communication using maps: an evaluation of their effectiveness. Bull Volcanol 70(2):123–138

Liu W, Dugar S, McCallum I, Thapa G, See L, Khadka P, Shakya P (2018) Integrated participatory and collaborative risk mapping for enhancing disaster resilience. ISPRS Int J Geo-Inf 7(2):68

Na J, Okada N, Fang L (2009) A collaborative action development approach to improving community disaster reduction by the Yonmenkaigi workshop method. J Nat Disaster Sci 30:57–69

Newig J, Kochskämper E, Challies E, Jager NW (2016) Exploring governance learning: how policymakers draw on evidence, experience and intuition in designing participatory flood risk planning. Environ Sci Policy 55:353–360

Okada N, Na JI, Fang L, Teratani A (2013) The Yonmenkaigi system method: an implementation-oriented group decision support approach. Group Decis Negot 22(1):53–67

Osti R, Tanaka S, Tokioka T (2008) Flood hazard mapping in developing countries: problems and prospects. Disaster Prevent Manage Int J 17(1):104–113

Parthasarathy D (2009) Social and environmental insecurities in Mumbai: towards a sociological perspective on vulnerability. South African Rev Sociol 40(1)

Pasupuleti R (2012) Towards an interdisciplinary framework for understanding the role of culture in the post disaster reconstruction process: the case of tsunami reconstruction in Tamilnadu, southern India. In: International disaster and risk conference: global risk forum, pp 565–568

Pasupuleti RS (2013) Designing culturally responsive built environments in post disaster contexts: tsunami affected fishing settlements in Tamilnadu, India. Int J Disaster Risk Reduct 6:28–39

Renn O, Webler T, Rakel H, Dienel P, Johnson B (1993) Public participation in decision making: a three-step procedure. Policy Sci 26(3):189–214

Reichel C, Frömming UU (2014) Participatory mapping of local disaster risk reduction knowledge: an example from Switzerland. Int J Disaster Risk Sci 5(1):41–54

Rowe G, Frewer LJ (2000) Public participation methods: a framework for evaluation. Sci Technol Human Values 25(1):3–29

Rowe G, Frewer LJ (2005) A typology of public engagement mechanisms. Sci Technol Human Values 30(2):251–290

Samaddar S, Okada N (2006) Participatory approach for post-earthquake reconstruction in the villages of Kachchh, India. Ann Disaster Prev Res Inst Kyoto Univ B 49:197–205

Samaddar S, Misra BA, Chatterjee R, Tatano H (2011) Participatory risk mapping for identifying spatial risks in flood prone slum areas, Mumbai. Ann Disaster Prevent Res Inst Kyoto Univ 54B:137–146

Samaddar S, Chatterjee R, Misra BA, Tatano H (2011b) Identifying vulnerability pattern in a flood prone micro-hotspot of Mumbai, India. In: 2011 2nd international conference on environmental science and development, IPCBEE, vol 4

Samaddar S, Misra BA, Chatterjee R, Tatano H (2012) Understanding community's evacuation intention development process in a flood prone micro-hotspot Mumbai. IDRiM J 2(2):89–107

Samaddar S, Misra BA, Tatano H (2012b) Flood risk awareness and preparedness: the role of trust in information sources. In: 2012 IEEE international conference on systems, man, and cybernetics (SMC). IEEE, pp 3099–3104

Samaddar S, Chatterjee R, Misra B, Tatano H (2014) Outcome-expectancy and self-efficacy: reasons or results of flood preparedness intention? Int J Disaster Risk Reduction 8:91–99

Samaddar S, Yokomatsu M, Dayour F, Oteng-Ababio M, Dzivenu T, Adams M, Ishikawa H (2015) Evaluating effective public participation in disaster management and climate change adaptation: insights from northern Ghana through a user-based approach. Risk Hazards Crisis Publ Pol 6(1):117–143

Samaddar S, Choi J, Misra BA, Tatano H (2015) Insights on social learning and collaborative action plan development for disaster risk reduction: practicing Yonmenkaigi system method (YSM) in flood-prone Mumbai. Nat Hazards 75(2):1531–1554

Samaddar S, Okada N, Choi J, Tatano H (2018) What constitutes successful participatory disaster risk management? Insights from post-earthquake reconstruction work in rural Gujarat. India. Nat Hazards 85(1):111–138

Samaddar S, Ayaribilla AJ, Oteng-Ababio M, Dayour F, Yokomatsu M (2019) Stakeholders' perceptions on effective community participation in climate change adaptation. In: Sustainable solutions for food security. Springer, Cham, pp 355–379

Samodra G, Chen G, Sartohadi J, Kasama K (2018) Generating landslide inventory by participatory mapping: an example in Purwosari Area, Yogyakarta, Java. Geomorphology 306:306–313
Tatano H, Samaddar S (2010) Flood risks reduction in livelihood risks: thoughts and insights from Mumbai. Ann Disaster Prevent Res Inst Kyoto Univ 53B:105–114
Tran P, Shaw R, Chantry G, Norton J (2009) GIS and local knowledge in disaster management: a case study of flood risk mapping in Viet Nam. Disasters 33(1):152–169
Webler T (1995) Right discourse in public participation: an evaluative yardstick. In: Renn O, Webler T, Wiedemann P (eds) Fairness and competence in public participation: evaluating models for environmental discourse. Kluwer, Dordrecht, The Netherlands, pp 35–86
White I, Kingston R, Barker A (2010) Participatory geographic information systems and public engagement within flood risk management. J Flood Risk Manage 3(4):337–346
Yadav M, Desai G (Year not mentioned) Transforming habitats: case study of Mumbai. Available at https://www.mumbaidp24seven.in/reference/30.pdf

Chapter 11
Grant for Global Sustainability Project: Enhancing the Urban Disaster Resilience of Kathmandu and Yangon Through Local Participatory Platform Activities

Glenn Fernandez and Kenji Okazaki

Abstract Catastrophic disasters have been occurring frequently all over the world in recent years, killing hundreds of thousands of people. Urban disaster risk is also on the rise due to rapid urban growth and due to vulnerable buildings and infrastructure. Hence, the enhancement of urban resilience against disasters is one of the urgent and important global issues. A Grant for Global Sustainability (GGS) Project was implemented by the Graduate School of Global Environmental Studies (GSGES) of Kyoto University in the pilot cities of Kathmandu, Nepal, Yangon, and Myanmar, aiming to enhance the resilience of the two cities against disasters through capacity building of local stakeholders. Kathmandu tackled earthquake risk while Yangon tackled earthquake, cyclone, and fire risks. The two cities established a local platform where stakeholders can work together to understand and assess the disaster risk of the city, estimate probable disaster damages, propose policies, and make action plans. In order to facilitate these activities, local counterpart organizations were selected from local governments, universities, and NGOs. In addition, the local universities and NGOs and GSGES conducted several joint research on urban seismic risk assessment; wind vulnerability; disaster education; risk perception and housing safety; and social fairness of action plans and policies to support the enhancement of urban disaster resilience. The goals of this project were to contribute to the 2030 Agenda for Sustainable Development, to attain effective and robust science–policy interfaces at the local level, and to contribute to the implementation of the Sendai Framework for Disaster Risk Reduction 2015–2030. The pilot cities are expected to transfer their experiences to other cities in their country and in neighboring countries. In order to disseminate the findings of the project and exchange information,

G. Fernandez (✉) · K. Okazaki
Graduate School of Global Environmental Studies, Kyoto University, Kyoto, Japan
e-mail: glenn@scu.edu.cn

K. Okazaki
e-mail: okazaki.kenji.28s@st.kyoto-u.ac.jp

Present Address:
G. Fernandez
Institute for Disaster Management and Reconstruction, Sichuan University, Chengdu, China

H. Tatano and A. Collins (eds.), *Proceedings of the 3rd Global Summit of Research Institutes for Disaster Risk Reduction*, Disaster and Risk Research: GADRI Book Series,
https://doi.org/10.1007/978-981-15-8662-0_11

international conferences were held twice, one at the beginning and another at the end of the three-year project.

Keywords Joint research · Science–policy interface · Sendai framework · Sustainable development

11.1 Introduction

Catastrophic disasters have been occurring frequently over the world in recent years, killing hundreds of thousands of people. Urban disaster risk is also on the rise due to rapid urban growth and vulnerable buildings and infrastructure. As the level of disaster resilience of a city can be defined as a total of physical vulnerability and capacity of all stakeholders, all the stakeholders should build up their capacity jointly to improve hard and soft measures in order to enhance the resilience of the city.

A Grant for Global Sustainability (GGS) Project was implemented by the Graduate School of Global Environmental Studies (GSGES) of Kyoto University. This GGS Project aimed to enhance resilience of cities against disasters through capacity buildings of stakeholders in a city. The pilot projects were conducted in two selected cities from Asia, which are prone to various severe disasters, namely Kathmandu (Nepal) and Yangon (Myanmar). Kathmandu targeted earthquake disasters, trying to assess ongoing recovery and reconstruction activities after the 2015 Gorkha Earthquake and to reflect the findings of the project to these activities to "Build Back Better." Yangon targeted fire, earthquake, and cyclone disasters. The project was conducted for nearly three years. It was expected that through the project activities all the stakeholders would be able to better understand the risks and probable damages caused by disasters as their own problems to be tackled and would be motivated or be willing to take appropriate actions for disaster reduction by themselves. In this way, the resilience of the pilot cities would be enhanced.

11.2 Project Implementation Methodology

Firstly, pilot projects were conducted in Kathmandu and Yangon. The two cities established a local platform in two pilot wards where stakeholders worked and discussed together to understand and assess the disaster risk of the city, estimate probable damages, propose policies, and make action plans. Some local projects for disaster education and community-based disaster risk management were also implemented. In order to facilitate these activities, local counterpart organizations were selected from local governments, universities, and NGOs. In Kathmandu, the counterpart organizations were: Lalitpur Metropolitan City (LMC); Center for Disaster Studies (CDS), Institute of Engineering (IOE), Tribhuvan University; and National Society for Earthquake Technology-Nepal (NSET). In Yangon, they were: Yangon

City Development Committee (YCDC); Faculty of Engineering, Yangon Technological University (YTU); and Myanmar Engineering Society (MES). A memorandum of understanding (MOU) was concluded between GSGES and the local counterpart organizations in the two cities to clarify the role and responsibilities of the participating parties. MOU with Kathmandu was concluded on January 22, 2016 and MOU with Yangon on February 12, 2016. The municipality managed the platform. The university and NGO gave technical advices to the platform and were assigned to implement projects for disaster education and community-based disaster management projects. The local counterpart organizations received technical support from Japanese experts and received financial assistance to cover part of actual expenses as a seed money. An international kick-off conference was organized in Kathmandu from January 21 to 23, 2016, and in Yangon from March 16 to 19, 2016, to launch the pilot project in the two cities and to initiate discussion on the establishment of local participatory platforms needed for the enhancement of urban disaster resilience and to conduct a field survey for the international participants. In addition to local platform activities, the local universities and NGOs and Japanese researchers conducted some joint researches to support the enhancement of urban disaster resilience. The joint research topics were: urban seismic risk assessment; wind vulnerability; disaster education; risk perception and housing safety; and social fairness of policies and action plans. The project timetable and project structure are shown in Fig. 11.1 and Fig. 11.2, respectively.

Secondly, a general model for urban resilience enhancement has been developed using a multidisciplinary approach, i.e., through a combination of engineering approach and political/social approach, based on the experiences of the pilot projects and the joint researches in the two cities. The model describes how to conduct similar projects to enhance the resilience against various disasters. The model is freely available with archived data on the project's dedicated Web site. The participating international organizations and experts will encourage other cities in the world to utilize

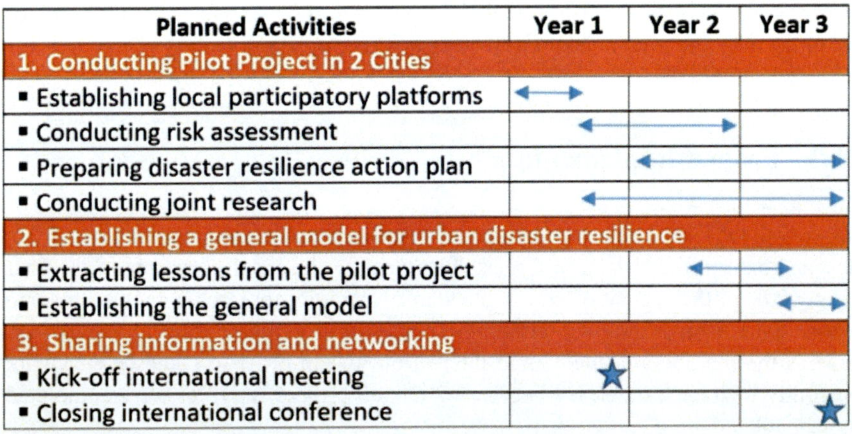

Fig. 11.1 Schedule of the activities

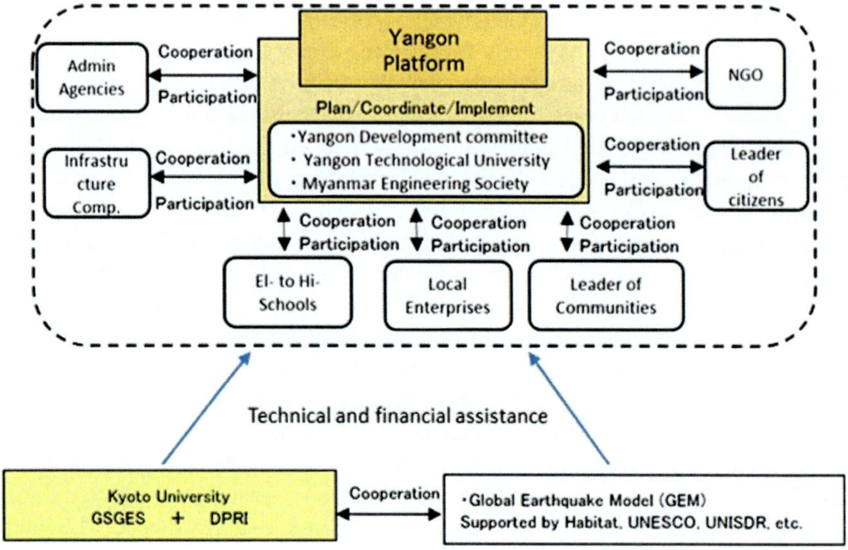

Fig. 11.2 Participating organizations and the established structure (Yangon)

the model through their international network. The pilot cities are also expected to transfer their experiences to other cities in the country and neighboring countries. In order to disseminate the findings of the project and exchange information, international conferences were held twice during the project, at the beginning and the end of the three-year project.

For risk assessment and damage estimate of earthquakes and capacity development, the Global Earthquake Model (GEM) Foundation assisted the two pilot cities at the start of the project. The GEM Foundation is a public–private partnership that drives a global collaborative effort to develop high-quality resources for assessment of earthquake risk and to facilitate their application for risk management. Some of the tools, data, and methods of GEM were applied to the pilot cities.

11.3 Project Implementation

In general, the GGS Project implementation in Kathmandu and Yangon was smooth despite some initial delays in the establishment of the local platforms in the four pilot wards in the two cities and in the start of the joint research activities of some teams. Establishing and maintaining local platforms and joint research teams took time as the members needed some interaction first in order to coalesce and work smoothly with each other. By February 2018, all project activities were completed as scheduled.

11.3.1 Local Platform Activities

Between January 2016 and February 2018, a total of 28 workshops was conducted under the GGS Project, 16 in Kathmandu, and 12 in Yangon. All the workshops were well attended, with the number of participants ranging from 35 to 108 per workshop. The cities made sure that different groups of stakeholders, including women, youth, and children, were represented in the workshops. In addition to the workshops, there were also five trainings (3 in Kathmandu and two in Yangon) conducted under the GGS Project. A total of 132 persons were able to obtain different kinds of training, contributing to the enhancement of their DRR-related knowledge and skills.

The Final Report of Lalitpur Metropolitan City (LMC) and of Yangon City Development Committee (YCDC) showing the details of the workshops and trainings that they conducted can be found on the project Web site, https://ggsurbanresilience.org/. The Urban Resilience Action Plan of the four pilot wards and the Final Report of the two NGOs supporting the local platforms, NSET in Nepal and MES in Yangon, are also posted on the Web site.

11.3.2 Joint Research Activities

During and after the kick-off conferences, the Japanese researchers and local experts discussed the research topics and elaborated on how the research should be conducted. They also identified the local joint research participants such as professors, students, and government officers. There were a total of eight joint research teams (five in Myanmar and three in Nepal) covering five research topics.

11.3.2.1 Urban Seismic Risk Assessment

The joint research team on urban seismic risk assessment focused on the vulnerability of existing masonry buildings in Kathmandu. The material properties of bricks and mortar, such as density, Young's module, Poisson's ratio, tensile strength, bond strength, friction angle, and compression strength, with and without the reinforcement, were obtained through the experiments. The loading tests of masonry walls with and without reinforcement were conducted for both in-plane and out-of-plane directions. The team also did the microtremor observations to grasp the characteristics of the ground in Kathmandu. The results can be used for the estimation of input ground motion to assess the risk of the structures. After obtaining expected earthquake ground motion, the collapse simulations using the obtained material properties were done by numerical simulation. In Yangon, the team investigated the damage patterns of local structure and characterized the damage grades to allow easy classification of damaged structures. By combining the seismic intensities with damage degrees due to past earthquake, fragility curves for local structure were constructed.

11.3.2.2 Wind Vulnerability

The joint research team on wind vulnerability wanted to bring attention to non-structural components of buildings such as roofing materials and fasteners. Wind-induced damages to non-structural components can be significantly reduced by improved construction and maintenance, such as using appropriate roofing materials with sufficient thickness, narrowing the intervals of fasteners, and replacing corroded roofing sheets. In their survey, team members used two methods: investigation by drone and direct visual investigation.

11.3.2.3 Disaster Education and Action Plan

The joint research team on disaster education and action plan applied two assessment tools, the Climate Disaster Resilience Index (CDRI) and School Disaster Resilience Assessment (SDRA), in the pilot wards in Kathmandu and Yangon. They investigated how to engage communities and schools in developing school-centered disaster resilience action plans.

11.3.2.4 Risk Perception and Housing Safety

The joint research team on risk perception and housing safety aimed to better understand the earthquake risk perception of people and their lifestyles in order to develop policies and strategies to motivate people to construct and maintain safer buildings. In Kathmandu, the team investigated how the houses and communities have been reconstructed after 2015 Gorkha Earthquake.

11.3.2.5 Social Fairness of Policies and Action Plans

The joint research team on social fairness of policies and action plans conducted a normative analysis of measurement problems concerning stakeholders' needs and studied justice in distributing goods and services among people who might have conflicting.

A Final Report has been prepared for each of the five research topics. These reports are also on the project Web site.

The Japanese researchers and international experts imparted their knowledge to the municipalities, NGOs, and universities in Kathmandu and Yangon. For example, Japanese researchers demonstrated to local counterparts how to properly conduct surveys and interviews and emphasized the importance of high-quality primary data. In the case of the joint research on risk perception and housing safety, the members of the Japan team checked the answers to the completed pre-survey questionnaires and provided feedback to the local university students on how to correctly record responses and how to elicit more detailed answers from the respondents through the use of examples or by asking follow-up questions. In terms of preparing house drawings, Japanese researchers showed the local university professors and students how to use the laser measuring device. They were also provided instructions and feedback on house drawings rendered in AutoCAD, to help them improve their proficiency in the use of the design software. In the other joint research topics, the local counterparts are exposed to the updated theories and methodologies being followed by the Japanese researchers. In the case of the joint research on wind vulnerability, a lecture on the background and theory behind the wind load modeling in the structural design code was conducted for the public by the team leader at the MES Building.

Briefing and debriefing meetings were frequently held by the joint research teams to explain the tasks to be carried out and to review if the tasks had indeed been performed properly. The local counterparts were given opportunities to share their knowledge and opinion or to ask questions.

The local counterparts mentioned that the joint research topics were new to them and that they were interested to be involved in the joint research because they wanted to learn more and because they believed that the topics would be useful in their own research and work.

The international experts also transferred knowledge to the local counterparts. By learning from GEM, NSET was already able to conduct the Resilience Performance Scorecard workshop by itself. What GEM may need to do next in the future is to help the NGOs and municipalities on how to analyze data and write a report. These skills will be helpful and useful to the NGOs and municipalities, not just for the GGS Project but also for their other current as well as future projects. GEM had also previously offered to provide online training to the local counterparts in Yangon on the use of the GEM tools in earthquake risk assessment.

Both Japanese and international experts shared their experiences from previous similar projects such as the Risk Assessment Tools for Diagnosis of Urban Areas against Seismic Disasters (RADIUS) initiative (Okazaki 2000) and from disaster risk management in Japan to the local counterparts during the workshops or in conversations. In return, the Japanese researchers and international experts were able to know more about the situation in Nepal and Myanmar from the local project partners. The GGS Project allowed all the participants to learn from each other.

At the end of the GGS Project, the local counterparts learned and understood well how to conduct and develop transdisciplinary joint research in several topics involving foreign and local universities, NGOs, local governments, and local residents. Their experience in the GGS Project will hopefully be useful to them when they initiate their own or participate in similar joint research projects in the future.

An international conference was held in Yangon on January 16–17, 2018, to mark the culmination of the three-year long GGS Project implemented in Myanmar and Nepal (GSGES-KU 2018). The two-day conference was coorganized by Yangon City Development Committee, Yangon Technological University, Myanmar Engineering Society, and the Graduate School of Global Environmental Studies of Kyoto University. Representatives from Lalitpur Metropolitan City, Nepal and researchers from Japan who were part of the project attended the conference.

On Day 1, representatives from Myanmar and Nepal presented the outcomes of their local platform activities and joint research activities to the 200 attendees at Yangon City Hall. Highlights included the presentation of the urban resilience action plans of the two pilot wards in Yangon, an open forum on how to move forward with action plan implementation, and a choreographed presentation by high school students about the disaster risk management cycle. Posters showing the outputs of different workshops in the past two years were displayed at the venue. The event was attended by the Mayor of Yangon and the Mayor of Lalitpur and was covered by six local media organizations.

On Day 2, a research symposium attended by 90 experts, practitioners, graduate students, ward-level stakeholders, and government officers was held at the MES Building. Representatives of the different joint research teams presented in detail their findings and recommendations. There were presentations on urban seismic risk assessment; wind vulnerability; disaster education; risk perception and housing safety; and social fairness of action plans. One professor and two Ph.D. students from YTU also gave a presentation related to urban resilience.

Based on the GGS Project implementation experience in Kathmandu and Yangon, a general model or guide for enhancing urban disaster resilience was prepared. There are step-by-step guides on how to replicate the local platform activities and joint research activities conducted under the GGS Project posted on the project Web site. It is expected that other cities will try to apply the model with assistance from international organizations. The pilot cities are also expected to transfer their experiences to other cities in their country.

11.4 Impact and Sustainability the Project

Through the collaboration among partners promoted by the GGS Project to establish the local platforms in the two pilot cities and through the various activities of the local platforms in the past two years, the stakeholders now have a better understanding the disaster risk in their communities. Through the risk assessments, scenario planning, action planning, and DRR-related training, it is expected that city and ward officials as well as ward residents now know what will be happening in case of a disaster and what kinds of actions should be taken to reduce disasters. Through the GGS Project, the stakeholders were able to develop an Action Plan and some of the activities in the Action Plan are already planned to be implemented. According to NSET, this is the first time in Nepal for scenario planning and action planning to be implemented at the ward level.

Impressed by the activities in Ward 16 and Ward 21, the Mayor of Lalitpur repeatedly promised to continue what the GGS Project had started. The experience of Ward 16 and Ward 21 will be replicated in the other 29 wards of Lalitpur using local funds. The Mayor of Yangon likewise mentioned during the international closing conference that the momentum created by the GGS Project will be sustained by continuously working the MES, YTU, and other stakeholders involved in the project to conduct similar activities in other townships. In the presentation of the YCDC focal person at the international closing conference, they intend to conduct additional assessments of existing vulnerable buildings in other townships similar to the rapid visual screening (RVS) survey conducted in Pazundaung and Tamwe under the GGS Project. During the courtesy call of the Mayor of Lalitpur to the Mayor of Yangon, both mayors discussed possible collaboration between the two cities. Yangon can learn many things related to disaster preparedness, emergency response, and disaster recovery from Lalitpur, which recently experienced the 2015 Gorkha Earthquake. The Mayor of Lalitpur extended an invitation to the Mayor of Yangon to visit Lalitpur in the near future.

During the final conference in Lalitpur, the last local platform activity under the GGS Project, the partner organizations in Nepal, LMC and NSET, toured the project leader and project researcher from GSGES, Kyoto University, around Ward 21 (February 22, 2018) and Ward 16 (February 23, 2018) to present the ongoing activities of the two pilot wards, which were part of the implementation of their Urban Resilience Action Plan. The group observed the removal of debris in different areas of Ward 21. Residents also showed their tools for emergency response. In the case of Ward 21, the tools were kept inside one of the schools. In Ward 16, the tools were on the ground floor of the ward office. The ward officials and residents expressed enthusiasm in continuing the activities started by the GGS Project. There are plans of procuring more equipment and supplies for their stockpile, as well as providing training to more volunteers in the ward. Everyone was pleased that they were involved in the GGS Project and could acquire knowledge and skills related to DRR.

In Yangon, the active and unwavering participation of school principals, teachers, and students was very noticeable right from the start of the GGS Project. The Disaster Education team was the first to conduct fieldwork in Yangon, even before the kick-off conference in that city. All activities to which the schools were invited always had a large number of participants. The principals and teachers in the pilot wards are very committed to educate their students about hazards and about disaster risk management. They also encourage their students to share what they have learned in school and in the GGS Project activities with their family members and with their friends and other schoolmates.

Because the principals in Yangon were themselves involved in the creation of the school-centered disaster risk management action plans (in addition to the ward's urban resilience action plan), there is a high probability that some of the activities in the action plans will be implemented. This is very important as schools are usually designated as evacuation areas or temporary shelters in Yangon. Given the central role of the schools in the action plans of the pilot wards in Yangon and the strong involvement of the principals, teachers, and students, the contribution of the GGS Project will have a long-lasting impact. The Mayor of Yangon has expressed interest in conducting activities similar to GGS Project activities in the other downtown wards of the city.

The outputs of the joint research teams in the form of journal articles (e.g., Fernandez et al. 2018; Furukawa et al. 2017); conference abstracts, presentations, and papers; and workshop presentations are also posted on the GGS Project Web site. Anyone interested in learning more about the research process can refer to the step-by-step guidance prepared by the different teams on how to conduct similar joint research projects in other cities.

This GGS Project aimed to contribute to the realization of the Sustainable Development Goals (SDGs), particularly Goal 4 (ensure inclusive and equitable quality education and promote lifelong learning opportunities for all), Goal 11 (make cities and human settlements inclusive, safe, resilient and sustainable), and Goal 17 (strengthen the means of implementation and revitalize the Global Partnership for Sustainable Development). Some of the local platform and joint research activities involved the promotion of safe and resilient housing through the participatory planning and management of human settlements with the aim of helping reduce the number of deaths and injuries due to earthquakes and cyclones. Through this GGS Project, organizations based in Japan and Italy assisted cities in Nepal and Myanmar in preparing disaster resilience action plans.

The GGS Project also aimed to contribute to the implementation of the Sendai Framework for Disaster Risk Reduction 2015–2030 (SFDRR) (UNISDR 2015), particularly in Priority 1 (understanding disaster risk) and Priority 2 (strengthening disaster risk governance to manage disaster risk). The GGS Project promoted the collection, analysis, and use of risk data and information, as well as facilitate the assessment of vulnerability, capacity, and exposure of the pilot wards in Kathmandu and Yangon. By working with the municipalities, the GGS Project helped to build the disaster risk management (DRM) capacity of government officials through sharing experiences, lessons learned, and good practices and through training and research. Local participatory platforms were established to promote and improve dialogue and cooperation among government officials, residents, NGOs, university researchers, and other stakeholders in order to facilitate effective decision-making in DRM.

The GGS Project likewise aimed to contribute to the "Human Security," which the Japanese Government is taking lead in promoting. Protecting human lives from disasters is one of the most important issues under "Human Security."

11.5 Future Plans

The pilot wards in the two cities have plans of implementing some of the activities in their urban resilience action plans. The Mayor of Lalitpur expressed his intention to replicate the experience of Ward 16 and Ward 21 to the other 29 wards of Lalitpur. Their first step will be to hold a training on community search and rescue to be facilitated by Ward 16. Three representatives from all the wards of LMC are expected to attend. NSET staff are expected to act as resource persons. Collaboration among the local government and NGO, which were strengthened by the GGS Project, will be able to promote such kind of activities even after the end of the GGS Project.

In Yangon, YCDC intends to conduct additional assessments of existing vulnerable buildings in other townships similar to the rapid visual screening (RVS) survey conducted in Pazundaung and Tamwe under the GGS Project. The Mayor of Yangon has expressed interest in conducting activities similar to GGS Project activities in the other downtown wards of the city.

Some of the joint research teams are in the process of preparing manuscripts for submission to scientific journals for publication and of preparing abstracts for submission to academic conferences.

Acknowledgements This chapter is a revised version of the interim project report (GSGES-KU and UNU-IAS, 2016). This GGS Project on urban disaster resilience (Contract No. ICA817), was financially supported by the United Nations University (UNU) with funds from the Ministry of Education, Culture, Sports, Science and Technology—Japan (MEXT).

References

Fernandez G, Tun AM, Okazaki K, Zaw SH, Kyaw K (2018) Factors influencing fire, earthquake, and cyclone risk perception in Yangon, Myanmar. Int J Disaster Risk Reduction. https://doi.org/10.1016/j.ijdrr.2018.02.028

Furukawa A, Kiyono J, Parajuli RR, Parajuli HR, Toki K (2017) Evaluation of damage to a historic masonry building in Nepal through comparison of dynamic characteristics before and after the 2015 Gorkha earthquake. Front Built Environ 3. https://doi.org/10.3389/fbuil.2017.00062

GSGES-KU (2018) International conference on urban disaster resilience held in Yangon. https://ggsurbanresilience.org/2018/01/17/international-conference-on-urban-disaster-resilience-held-in-yangon/

GSGES-KU & UNU-IAS (2016) Enhancement of urban disaster resilience through activities of local participatory platform (interim report). Kyoto, Japan: Graduate School of Global Environmental Studies, Kyoto University & United Nations University—Institute for the Advanced Study of Sustainability. https://i.unu.edu/media/ias.unu.edu-jp/news/10923/GGS-progress-report-2016_Kyoto-University.pdf

Okazaki K (2000) RADIUS: risk assessment tools for diagnosis of urban areas against seismic disasters. UNISDR, Geneva, Switzerland. https://www.unisdr.org/files/2752_RADIUSRiskAssessment.pdf

United Nations International Strategy for Disaster Reduction (UNISDR) (2015) Sendai Framework for Disaster Risk Reduction 2015–2030. https://www.wcdrr.org/uploads/Sendai_Framework_for_Disaster_Risk_Reduction_2015-2030.pdf

Chapter 12
Seismic Vulnerability Assessment of Existing RC Buildings in Ranau, Sabah Malaysia

Mariyana Aida Ab Kadir, Mohd Zamri Ramli, Azlan Adnan, and Mohd Nur Asmawisham Alel

Abstract Malaysia is located in a low seismicity area, thus the buildings in Malaysia were not designed as seismic resistant structures. However, on 5 June 2015, Malaysian was been shocked by the 5.9M Ranau Earthquake which killed 18 climbers of 5 nationalities and was recorded as the largest earthquake in Malaysia since 1976. The earthquake affected some buildings in the nearby area. The damaged associated with these events reflected the construction practice in Malaysia. The low awareness among the general public towards structural safety and the inability of regulatory bodies and technical professionals in maintaining quality standards in constructions has created an urgent need to educate the leaders, public, city planners, architects and the engineering professionals about the consequences of earthquakes. As a step in understanding the seismic risk to buildings in Malaysia, there is a need to determine the vulnerability of existing buildings, particularly in Ranau, Sabah, as this region has been identified as high seismicity area in Malaysia. When classes of buildings are considered for risk assessment, the vulnerability can be established in terms of their structural characteristics, and suitable modifiers to the vulnerability function can be established in terms of their geometrical characteristics. In this study, the seismic capacity of the selected existing RC buildings has been evaluated through nonlinear dynamic simulations. Seismic response has been analysed, considering various peaks and integral intensity measures and various response parameters, such as ductility. A structural performance and damage level for each studied type are presented.

Keywords RC buildings · Earthquake affected buildings · Risk assessment · Numerical analysis

---

M. A. A. Kadir (✉) · M. Z. Ramli · A. Adnan · M. N. A. Alel
Department of Structure and Materials, School of Civil Engineering, Faculty of Engineering, Universiti Teknologi, Johor Bahru, Malaysia
e-mail: mariyanaida@utm.my

H. Tatano and A. Collins (eds.), *Proceedings of the 3rd Global Summit of Research Institutes for Disaster Risk Reduction*, Disaster and Risk Research: GADRI Book Series,
https://doi.org/10.1007/978-981-15-8662-0_12

12.1 Introduction

On Friday, 5 June 2015, a 5.9 Richter scale earthquake rocked Sabah as shown in Fig. 1.1. The earthquake, with a depth of 10 km, occurred at 6.1°N 116.6°E, 16 km from Ranau, Sabah and the tremors were felt throughout Sabah, particularly Kundasang, Tambunan, Pedalaman, Tuaran, Kota Kinabalu and Kota Belud (Felix Tongkul 2016). The location and intensity of at the Ranau area has been predicted in the Seismic Hazard Map established in 2005 (Adnan et al. 2005; Harith et al. 2015)

The earthquake also shook Mount Kinabalu, Sabah the highest mountain in Malaysia. A total of 18 people of 5 nationalities (Singaporean, Malaysian, Chinese, Japanese and Filipino) was confirmed dead. The victims, climbers and a mountain guide died due to severe injuries caused by massive falling rocks and boulders induced by the ground motion. The 5.9M Ranau Earthquake was recorded as causing the highest number of casualties and greatest structural damage in Malaysian history. Cracks were reported in residential buildings, commercial buildings, resorts and hotels as well as places of worship. Even infrastructures, which are normally employed in an emergency, such as the hospital, schools and police department were not spared. Figure 1.2 shows the damage found on some buildings. The most serious damage occurred in the rest house near the summit of Mount Kinabalu (Fig. 1.2b) and buildings in Kota Belut and Tuaran were similarly affected by the earthquake. Typical types of damage reported were brick wall shear failure cracks, cracks in columns and beams, roof failure, failure of supporting columns or tilts, concrete spalling and shattered windows (ATC, 1996 and American Society of Civil Engineer 2013).

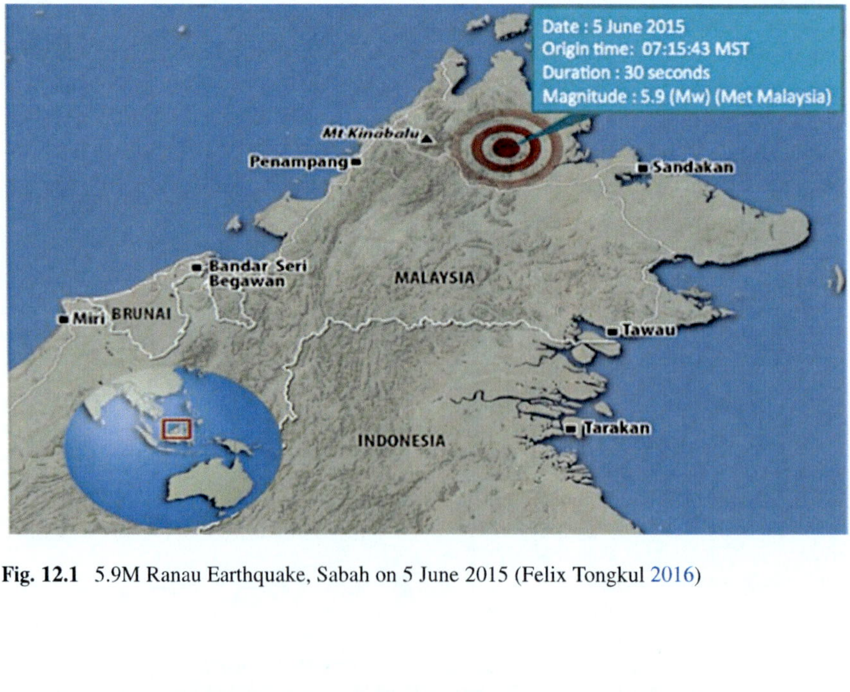

Fig. 12.1 5.9M Ranau Earthquake, Sabah on 5 June 2015 (Felix Tongkul 2016)

(a) (b)

Fig. 12.2 Damage to buildings due to Ranau Earthquake **a** shear failure at residential, **b** column failure at rest house

There were worries about whether the continuous swaying of buildings and vibration of some parts of the non-structural elements, such as doors and windows, would happen continuously and affect the integrity of the whole buildings. Therefore, the objectives of this study are to simulate the earthquake loads acting on the existing reinforced concrete buildings in the Ranau area and study their effects through finite element modelling to develop demand–capacity curves in order to investigate the vulnerability of the building due to earthquake motion.

12.2 Numerical Analysis

Nonlinear analysis using the ETABS building analysis and design program, SAP2000 Integrated Structural Analysis and Design Software, was used to conduct the study. In addition, the model also performed the demand and capacity analyses (DCA) (Fardis et al. 2005). Three existing reinforced concrete buildings in the Ranau area were modelled to represent the building main frames, according to their structural drawings and material properties. These were: Building 1; Asrama Perempuan, SMK Mat Salleh, Building 2; Asrama Lelaki, SMK Mat Salleh and Building 3; Kompleks Pentadbiran Kerajaan Persekutuan.

The earthquake loads were then statically combined with the dead and live loads to estimate the structural adequacy of the building (Copra and Goel 1999). Response spectrum analysis was implemented to study the performance of the building subjected to earthquake. The model was analysed nonlinearly by applying a series of seismic loadings to determine the failure mechanism. Nonlinear static (pushover) analysis was carried out to estimate the strength, ductility and expected performance of the designed buildings in finite element modelling (FEM) analysis. The accuracy of pushover analysis depends on a number of factors including the distribution of the lateral load and type of buildings. The procedure used to obtain

Table 12.1 Mode shapes

No.	Building name	Mode shapes (sec) and directions			
		1	2	3	4
1	Asrama SMK Mat Salleh—Asrama Perempuan Ranau	0.60y	0.59t	0.39x	0.24t
2	Asrama SMK Mat Salleh—Asrama Lelaki Ranau	0.87y	0.86t	0.70x	0.29t
3	Kompleks Pentadbiran Kerajaan Persekutuan, Ranau	0.84x	0.79y	0.70r	0.32t

Note x = x translational direction; y = y translational direction, r = random, t = torsional

the performance point was represented by the expected peak displacement of the structure, the point of intersection between the capacity spectrum and demand spectrum. The analysis was able to observe the earthquake-resistant design and detailing by monitoring the strength and ductility capacity of the building.

12.3 Results and Discussion

12.3.1 Free Vibration Analysis

The free vibration study concluded that the dynamic characteristics of the finite element models are acceptable when compared with the equations in Euro Code 8 (EuroCode8) and the International Building Code (IBC). The summarized values of the first four mode shapes are listed in Table 3.1 and are used for verifying the dynamic behaviour of the models and their suitability for further analysis.

12.3.2 Linear Analysis

12.3.2.1 Base Shear Force of Buildings

The performance of buildings in linear capacity was analysed according to Eurocode 8: (1) static lateral force analysis method (SLA) and (2) response spectra analysis method (RSA). The intention of these analyses at the linear stage was to verify the adequacy of the beams' and columns' capacity against the internal forces developed under both static and dynamic analyses. The percentage of building weights is mostly higher from previous study due to the high recommended PGA values of 0.14 and 0.18 g.

Table 12.2 Demand–capacity

No.	Building name	Demand–capacity	
		2D	3D
1	Asrama SMK Mat Salleh—Asrama Perempuan Ranau	OP	IO
2	Asrama SMK Mat Salleh—Asrama Lelaki Ranau	OP	IO
3	Kompleks Pentadbiran Kerajaan Persekutuan, Ranau	OP	OP

Generally, the results from response spectrum analysis show higher values in moment and axial forces for columns, in comparison with lateral load analysis. The results for the beams also show a similar trend. These phenomena could be due to the higher contribution of the mode shapes in the responses for complicated structures, which is not captured in the lateral load analysis. It indicates that the structural members may be cracked and damaged under earthquake loading.

Linear static and dynamic analyses were performed in this study. The buildings were identified to be sufficient to take earthquake forces, based on the locations of the buildings in Sabah. In general, columns are performed under the allowable limits of bending and shear. Nevertheless, the beams of the buildings require stronger sections to take bending forces from the earthquake loads.

12.3.3 Nonlinear Analysis

From the demand–capacity results shown in Table 3.2, it can be concluded that there are no indications that major cracks will occur. However, minor cracks will occur to almost all buildings except for Building No. 3.

12.4 Conclusion

For an earthquake intensity level in Sabah of about 0.20 g, it is predicted that buildings will not have any problem with the seismic load from the six time-histories used for the analyses. They are all within OP level.

12.5 Recommendations

Further analyses need to be performed, especially to study the behaviour of retrofitted buildings and compared with the current conditions. In summary, buildings with developed plastic hinge rotations at the bottom of columns at ground level are proposed to be installed with bracings or dampers at the building bottom levels, in their weak frame directions only.

Acknowledgements The author would like to thank the HICOE Research Grant (4J223), Universiti Teknologi Malaysia and Aun/SEED for financial research grant support.

References

Applied Technology Council, ATC (1996) Seismic evaluation and retrofit of concrete buildings (ATC40). Appl Technol Council (ATC) 1

American Society of Civil Engineers (2013) Seismic evaluation and retrofit of existing buildings, American Society of Civil Engineers

Adnan A, Hendriyawan, Marto A, Selvanayagam B (2008) Development of seismic hazard maps of east Malaysia. In: Advance in earthquake engineering applications. Book Chapter, Penerbit UTM

Chopra AK, Goel RK (1999) Capacity-demand-diagram methods based on inelastic design spectrum. Earthq Spectra 15(4)

Eurocode 8: design of structures for earthquake resistance, part 1: general rules, seismic actions and rules for buildings (2013), The British Standards Institution

Fardis MN, Carvalho E, Elnashai AS et al (2005) Designer's guide to EN 1998–1 and EN 1998–5. Eurocode 8: design of structures for earthquake resistance, general rules, seismic actions and design rules for buildings, foundation and retaining structures. Thomas Telford Ltd., London

Harith NSH, Adnan A, Shoushtari AV (2015) Seismic hazard assessment of east Malaysia region. In: International conference on earthquake engineering and seismology (IZIIS-50), 12–16 Mei 2015, Floating MSC ORCHESTRA cruiser

Tongkul F (2016), The ranau earthquake 2015 cause and impact, Sabah Soc J 1–28

Chapter 13
Large Landslides in México in the Past 10 Years

Patricia Alarcón, José Beltrán, and Jaime Verduzco

Abstract Due to its geographical location, Mexico is influenced by considerable natural hazards such as earthquakes, hurricanes, and landslides, among others. The states of Guerrero and Michoacán have been scenarios of large landslides like the ones of Cerro El Socorro at La Huahua, Michoacán, in October 2008; seven landslides in Mineral de Angangueo, Michoacán, in February 2010; and the landslides of La Pintada, Guerrero, in 2013—all of these linked to meteorological phenomena—causing damage to infrastructure as well as deaths, and in particular the wiping out of the population of La Pintada in Guerrero. This presentation outlines the evaluation of landslides through geophysical, seismic refraction, and geotechnical techniques. An evaluation of the economic and social vulnerability of the inhabitants of the three affected communities will also be presented.

Keywords Landslides · Earthquake · Slides · Floods · Early warning systems

13.1 Landslide of La Huahua, Aquila, Michoacán, México

The objective of this study is to analyze the risk management in the presence of a trigger event and the probable sequential occurrence of one or two environmental phenomena that could magnify the risk potential. The linked environmental threat produced in limited time and space causes severe unforeseeable damage whose periodicity is almost impossible to anticipate. A case for study which occurred on the Mexican Pacific coast is presented, in which the occurrence of an earthquake activated a movement of land masses which in turn generated river stagnation and floods. Finally, the development of instruments and concrete actions of intervention for the evaluation, risk control and disaster reduction in the face of this scenario is proposed,

P. Alarcón (✉)
Structures Laboratory, Civil Engineering University of Michoacán, Morelia, México
e-mail: riskplanet@gmail.com

J. Beltrán · J. Verduzco
Research Institute for Risk Management. Morelia, Michoacán, México

H. Tatano and A. Collins (eds.), *Proceedings of the 3rd Global Summit of Research Institutes for Disaster Risk Reduction*, Disaster and Risk Research: GADRI Book Series,
https://doi.org/10.1007/978-981-15-8662-0_13

with the goal to establish a conceptual and operative frame which will allow locating and measuring the actions, and particularly the tools, for corresponding intervention.

13.1.1 Linked Threat Earthquake-Slide-Floods

According to the Institute of Geophysics at UNAM, an earthquake was registered on September 26, 2008. On Wednesday, October 1, inhabitants of the village of La Huaha, located on the coastal highway between Manzanillo and the port of Lázaro Cárdenas (Fig. 13.1), reported the presence of 500-m-long cracks along the crest of Socorro Hill on the right bank of the Huahua River, approximately 8 kms upstream of said village. On October 4, the hillside collapsed. This hill is located far enough from any human settlement; however, the closest one, La Ujera, is only 2 kms away, reason why the residents were evacuated from their homes due to a potential flood risk. The potential risks appeared after the rockslide, as the displacement of the hillside was of extraordinary dimensions. As a result of the slide, terraced cuts on the hillside became exposed. (The displaced land masses formed a dam (plug) on the preexistent faults on the river over a length of 1100 m and a height of 800 m, raising the level of the water upstream to between 35 and 40 m.

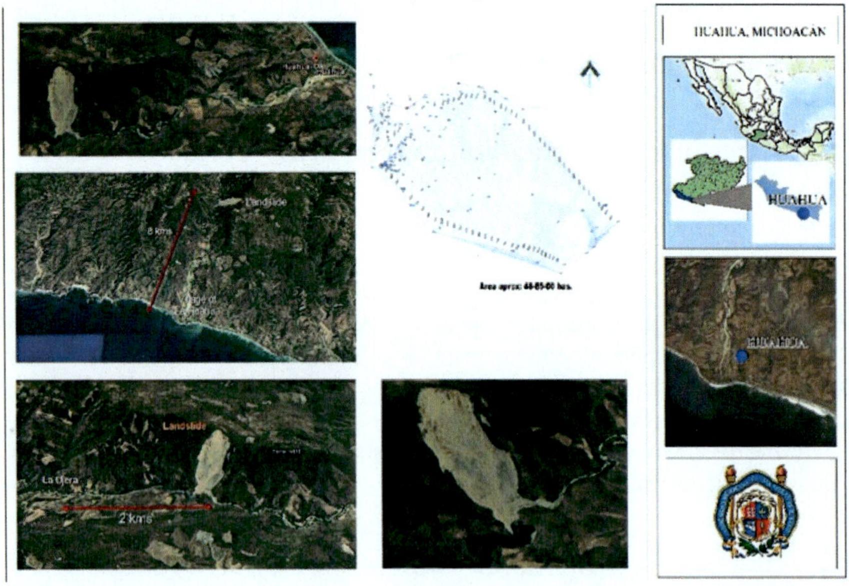

Fig. 13.1 Located of the village of la Huahua and landslide

13.1.2 Landslide

The rockslides on the left flank of Socorro Hill are located on the right bank of the Huahua River, approximately eight kms upstream of the village of La Huahua. The area belongs to the so-called Guerrero—Colima Orogenic Complex Province, of the Mesozoic with a Laramidic age of between 90 and 40 million years. Moreover, it is located over huge batholiths of granitic composition from the Superior Cretaceous, which presents superficial filling deposits and/or sandy soil of scant thickness.

Four layers or rock bodies were read on the cross-sections or electro-stratigraphic profiles (Fig. 13.1), which are recorded as follows:

1. The first of them, and the most superficial, is constituted of strongly weathered granites with a thickness of 200 m.
2. The second suddenly increased toward the SE to reach 570 m, with electric resistances from 232 to 489 Ω m (blue colors), which indicate its altered state and bad rock quality. Obviously, this package is likely to slide in the future (Fig. 13.2).
3. Here are underlying granites partially fractured and altered, with an average thickness of 400 m, and electric resistances from 514 to 1792 Ω m (green color), which indicate good rock quality guaranteeing stability of the rocky massif.
4. Finally, underlying the previous package and with an undefined thickness, we can appreciate electric resistances from 3279 to 6000 Ω m (yellow, orange and red colors) which correspond to the healthy granites which constitute the underlying rock of the site.

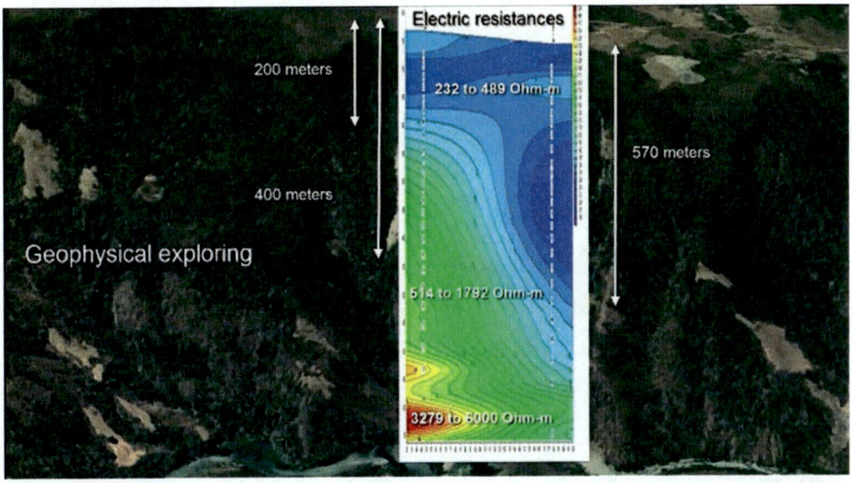

Fig. 13.2 Cross sections or electro-stratigrafic profiles, of landslide of Socorro Hill, La Huahua, Michoacán, México

13.1.3 Floods

The detached material that slid into the La Huahua River formed an obstruction or dam. According to instrumental seismic refraction exploration realized during the following days, it was possible to determine that losses of hydraulic charge would happen. With this precedent, it was noted that an increase of rainfall would lead to the collapse of the dam, resulting in an avalanche of sandy material and a tsunami-like wave that would affect the community of La Huahua (Fig. 13.3). The maximum height of the dam holding in the water masses was approximately $h = 35$ m. With these data, the following parameters were estimated; The accumulated water volume in the reservoir was approximately; 7′832,325m^3; Cumecs $Q = 3.5$ m^3/s; Surface flooded of $S = 385{,}690$ m$^{2\cdot}$; The volume of the detached material is 11′253,583 m^3, a value correspondent to the dam section at the base of the slope, in great part of clay and sand-slime material which expanded over the river.

Once having identified the hazards through various geophysical and analytical techniques, it was proposed to study the zone employing a risk management system in order to reduce its vulnerability.

The case study above illustrates an example of how a quake with an epicenter 150 kms distant caused a landslide, flooding, and a significant change in landscape geology.

The rural communities in the surroundings of the area are still threatened by the active movement of El Socorro Hill, high seismic activity in the region, and

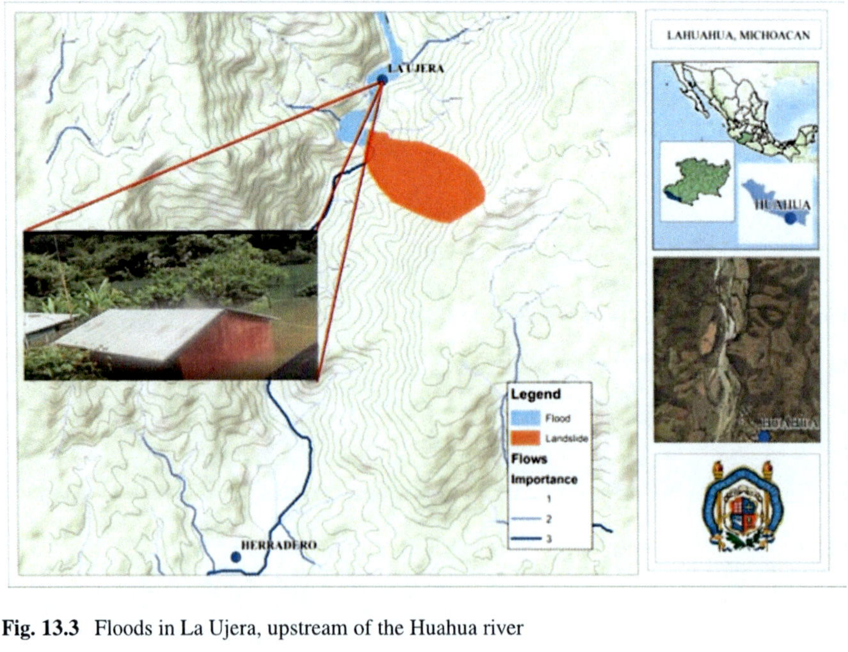

Fig. 13.3 Floods in La Ujera, upstream of the Huahua river

the recurrent presence of hurricanes and tropical storms. Any of these threats could produce future landslides on El Socorro Hill, and, as a result, the dimensions of the dam formed by detached material would increase leading to flooding which would affect the rural communities upstream.

The conditions described above have not been intervened properly; neither has there been any installation of early warning systems due to a lack of economic resources and a lack of action by authorities. The rural communities have adapted to this new condition, taking advantage of the dam formed by the detached material which created a natural scenery of exotic beauty, such as a natural water reservoir in the middle of the jungle, with huge granite boulders and a slope in green, orange and white colors, which at the same time presents the birth of a new natural environmental order. The locals who used to fish now offer tourist rides on and around this reservoir.

It is urgent to intervene in this case study scenario. The high social and economic vulnerability of these rural communities may result in a lack of conscience about the risk levels present in the zone, as they adapt to the changes in an attempt to cope with this new situation, thereby creating a risk cycle.

13.2 Mineral de Angangueo, Michoacán, México

During the first days of February 2010, heavy rains throughout most of the country led to multiple threats in the town of Mineral de Angangueo, (Fig. 13.4) a mining

Fig. 13.4 Located of Mineral de Angangueo and the most import landslide

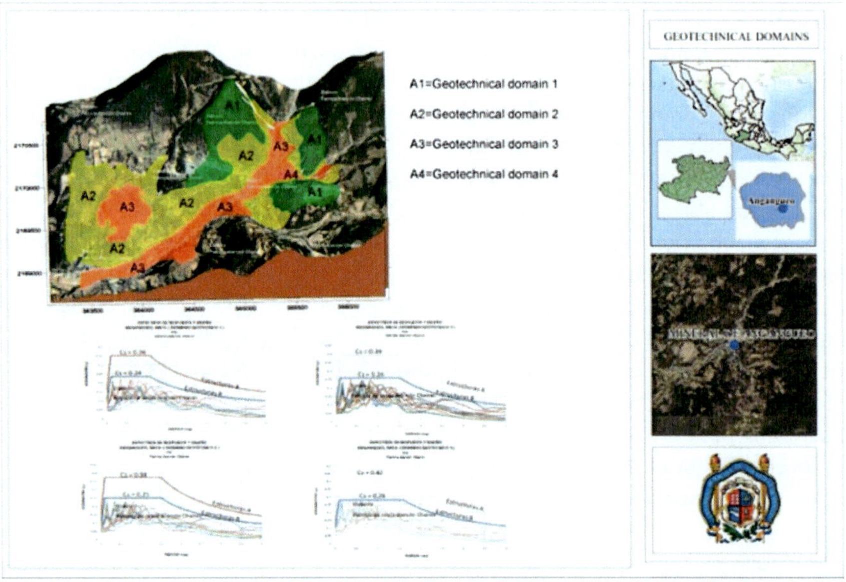

Fig. 13.5 Example of hazard map, seismic microzoning map

community in the state of Michoacán, among which were the overflowing and the recognition of the former bed of the main river, landslides, the blockage of the main river, and flooding. These caused a significant loss of lives as well as severe damage to infrastructure, homes, and public buildings.

During the evaluation of the landslides, it was observed that the saturation of the clay and silt soils which form the surface layer of the lithology of the site, as well as the steep slopes, deforestation and the impact of human activity (mining) led to the loss of soil resistance in the rocky hillsides that surround the mining town, leading to major landslides (Fig. 13.5). The proposal of the University of Michoacán (San Nicolas de Hidalgo) and the Research Institute for Risk Management (an NGO) to the local government was the implementation of a public policy to manage the disaster through the priority actions of the Hyogo Framework.

The hazard identification was carried out by initially considering the history of the exploitation of natural resources and its effect on disastrous events in the local and collective memory. In addition, technical and scientific exploration (in geophysics, soil mechanics, the modeling of hazard scenarios of unstable slopes, seismic, hydrological, and anthropogenic) was conducted which allowed the elaboration of hazard maps and buffer zones for land use planning (Fig. 13.6).

Local women and youngsters have been involved in addressing the damage, and they have positively influenced the rest of the community. Currently work is being continued to increase community resilience. In this sense, it is important to note that in order to perform these tasks, we have found as root vulnerabilities dynamic pressure and unsafe latent conditions that prevent DRR and risk management (RM) to

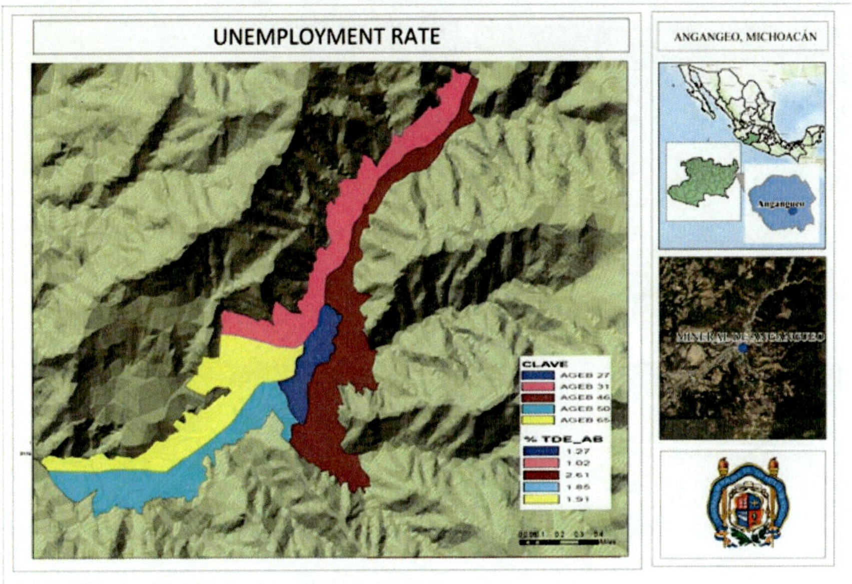

Fig. 13.6 Example of social vulnerability map

be effectively consolidated. Finally, as an University we consider that the experience of disaster and risk management in Mineral de Angangueo, Michoacán is a good start for the consolidation of the priority guidelines of the Hyogo Framework and Sendai Framework in Mexico. However, it is important to recognize that there is much more work to be done with the affected community which, throughout centuries, has been economically and socially marginalized by the mining activity. On the other hand, there must be better synchronized action between the different agents (government, technical-scientific sector and the community in general) for effective risk management in order to establish a new and better understand ding between mankind and nature.

13.3 La Pintada, Atoyac de Álvarez

In September 2013, the almost simultaneous occurrence of two severe meteorological events—tropical storms *Ingrid* and *Manuel*—triggered and even more extreme meteorological phenomenon that ravaged the Mexican states of Guerrero, Veracruz, Hidalgo, Oaxaca, Jalisco, Michoacán, Sinaloa, Puebla, and Tamaulipas. As a result of these phenomena, there were reports of approximately 157 deaths, 1.2 million affected victims, 39,000 evacuees, and 77 municipalities officially declared disaster zones. Among these devastating effects, the village of La Pintada, located in the Municipality of Atoyac de Álvarez in the western state of Guerrero, was completely

Fig. 13.7 After, during, and before of landslide

buried by a huge landslide. The region's geological context is known as *Terreno Guerrero* and is located within the Zihuatanejo-Huetamo Subterranean, according to the division proposed by Campa (1978). Researchers explored the affected area in general and examined the surface of the landslide in order to observe and identify its characteristics onsite (Fig. 13.7).

The region's geological context is known as *Terreno Guerrero* and is located within the Zihuatanejo-Huetamo Subterranean, according to the division proposed by Campa (1978). Researchers explored the affected area in general and examined the surface of the landslide in order to observe and identify its characteristics onsite. The classification of the soil was MH (inorganic slit). The landslide was set off by a concatenation of various natural phenomena and several kinds of natural hazards. It also mentions some natural antecedents which acted as detonators of the massive landslide that buried the community.

The issue of social vulnerability must be analyzed in detail in order to come up with pragmatic solutions that will bring about public policies as well as economic and social development in the State of Guerrero, Mexico. It is high time not only to reflect but also to recognize that society is the foremost creator of risks in the presence of extreme or not extreme events, due to the lack of land-use planning, poverty, inequality, corruption, and a deficient government. In Fig. 13.8, we can see that, it is on this, thin line, where life and societies constructing new risk are located again in La Pintada Guerrero.

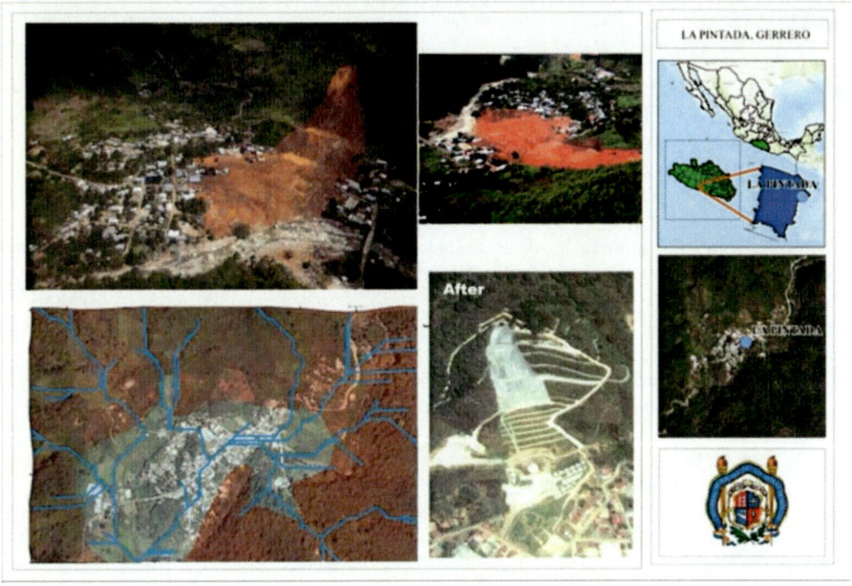

Fig. 13.8 Thin line, where life and societies constructing new risk are located again in La Pintada Guerrero

13.4 Conclusions

1. A great number of disasters have occurred as a result of several linked threats.
2. Chaos seems to be part of nature's structure and is linked to self-regulating events; in this sense, it is proposed to explore models of analysis and algorithms based on complex systems, which may allow the development of mitigation processes and early warning systems to reduce vulnerability.
3. For this, we would have to raise our capability of analysis and understand that Earth is a complex system which tends to regulate itself, which is shown in the linked threats producing new transitory states or states of entropic rest (Ilya 1977).
4. The interaction between nature and human being as a part of it tends to search for a natural order, staying inside the tight domain oscillating between unchangeable order and total chaos, between a rigid constancy and anarchic turbulence, a condition on the edge of chaos (Langton 1990).
5. The interaction between nature and human being as a part of it tends to search for a natural order, staying inside the tight domain oscillating between unchangeable order and total chaos, between a rigid constancy and anarchic turbulence, a condition on the edge of chaos (Langton 1990).

References

Alarcón P (2008) Informe Técnico de Gestión de riesgos del deslizamiento e inundación en el Cerro el Socorro en la Huahua, Municipio de Aquila, Michoacán, México. Dirección Estatal de Protección Civil. Michoacán, México

Alarcón P (2010a) Tesis de Maestría en Construcción. Sistema de gestión integral del riesgo para la planificación y administración de las obras de ingeniería. Instituto Mexicano de la Industria de la Construcción, México

Alarcón P (2010b) Evaluación de los deslizamientos e inundaciones en Mineral de Angangueo en 2010. Dirección Estatal de Protección Civil. Michoacán, México

Alarcón P (2011) Gestión del Integral del Riesgo para el ordenamiento territorial en Mineral de Angangueo, Michoacán. Universidad Michoacana, México, Tesis para obtener el grado de Doctor en Ciencias

Alarcón P (2013) Evaluación del deslizamiento en la Pintada Guerrero por eventos meteorológicos de Ingrid y Manuel. Dirección Estatal de Protección Civil, Municipio de Atoyac de Álvarez, Guerrero, México

Alianza Clima y Desarrollo (2012) La Gestión de Riesgos de Eventos Extremos y Desastres en América Latina y el Caribe: Aprendizajes del Informe Especial (SREX) del IPCC

Campa UMF (1978) La Evolución Tectónica de Tierra Caliente. Guerrero, Boletín de la Sociedad Geológico-Mexicana, Tomo XXXIX

Ilya P (1977) Tan solo una ilusión. Prigogine, Una exploración del caos al orden, 4ª edición, Trusquets Editores, España

Porraz ZR (1962) Investigaciones Mineras de la Cuenca del Río Balsas. C.R.M, Oficina Regional Chilpancingo

Chapter 14
Disaster Risk Reduction (DRR) and Management Under a Changing Climate: Bridging the Divide Through Integrated Coastal Management (ICM) Toward Sustainable Development Goals

Johanna Diwa-Acallar

Abstract Climate change poses a grave threat in coastal areas which have already started feeling its impacts including flooding, increasing frequency and strength of typhoons and tropical storms, sea level rise, storm surges, etc. A clear strategy and action program is vital as countries progress toward their respective sustainable development objectives. Through Integrated Coastal Management (ICM) development and implementation in the region, countries in East Asia are progressing toward parallel Sustainable Development Goals (SDGs) especially in objectives related to poverty, food security, economic progress and ecosystems conservation. Specific issues brought about by climate change like flooding, storm surges, sea level rise, disasters, pollution among others are also being addressed through management mechanisms and action plans put in place in their respective national development plans.

The Sustainable Development Strategy for the Seas of East Asia (SDS-SEA), a regional framework among country and non-country partners and collaborators, serves as a platform for regional collaboration under the framework of the Partnerships in Environmental Management for the Seas of East Asia (PEMSEA). A partnership arrangement specializing in integrated coastal and ocean governance of the Seas of East Asia, PEMSEA has 11 country and 20 non-country partners, all committed to the implementation of the SDS-SEA plan.

In this paper, two case studies are presented to demonstrate how the Sustainable Development of Coastal Areas (SDCA) framework can address various aspects of Sustainable Development on Habitat Protection Restoration and Management, Food Security and Livelihood Management, Pollution Reduction and Waste Management and the Natural and Human Induced Hazard Prevention and Management.

Da Nang, a city in the coastal area of Vietnam, is one of PEMSEA's ICM sites that demonstrates how ICM projects are providing learning experiences at the local

J. Diwa-Acallar (✉)
Capacity Development Manager, Partnerships in Environmental Management for the Seas of East Asia (PEMSEA), Quezon City, Philippines
e-mail: joandiwa@hotmail.com

H. Tatano and A. Collins (eds.), *Proceedings of the 3rd Global Summit of Research Institutes for Disaster Risk Reduction*, Disaster and Risk Research: GADRI Book Series,
https://doi.org/10.1007/978-981-15-8662-0_14

level, building capacities in hazard and vulnerability mapping and land and sea-use zoning, enhancing forecasting and early warning systems, natural and man-made defenses and emergency response and decreasing vulnerability, particularly of marginalized and impoverished coastal communities. With the evident impacts of climate change in the city, the implementation of ICM has provided a framework for comprehensive planning and management that enable its government to address disaster risk reduction through integrated governance programs. Since its adoption of ICM in 2000, Da Nang City has successfully overcome some of the weaknesses in its governance system and the limitations in the coordination and planning processes through integrated action plans to address climate change.

Guimaras, which is an island province located in the middle part of the Philippines, is also prone to natural disasters such as storm surges, flood, landslides, and soil erosion due to the impacts of climate change. Following the SDCA Framework, the Guimaras embraced ICM as main vehicle of its many programs and projects especially for sustainable development of its coastal areas through integrated management approach in implementing Ecosystem Conservation of Critical Habitat.

Keywords Climate change · Disaster risk reduction · Integrated coastal management · Regional ocean governance · Strategic framework · Sustainable development goals

14.1 Introduction

Climate change poses a grave threat in coastal areas which have already started feeling its impacts including flooding, increasing frequency and strength of typhoons and tropical storms, sea level rise, storm surges, etc. In spite of these threats, countries are expected to move toward their Sustainable Development Goals, making sure their national policy, strategies, and guidelines are in place to respond to these challenges through implementation of appropriate preventive and management measures.

Strategic action programs in climate change adaptation and disaster risk reduction are vital now than ever as countries progress toward their respective sustainable development objectives. Through Integrated Coastal Management (ICM) development and implementation, countries in East Asia are progressing toward parallel Sustainable Development Goals (SDGs) especially in objectives related to poverty, food security, economic progress and ecosystems conservation.

The existing synergies and their principles, tools, strategies as well as governance aspects that comprise the ICM framework are very apparent between the SDGs and the SDS-SEA, such as SDG 14: Life Below Water, SDG 6: Clean Water and Sanitation, SDG 11: Sustainable Cities and Communities, SDG 17: Partnerships for the Goals, and specifically SDG13: Climate Actions. Through 'Adapt' strategy, a main feature in the updated SDS-SEA, CCA/DRR and management are incorporated into development policies, plans, and programs at national, provincial, and local levels.

14.2 Background

The threats of climate change if not addressed adequately could lead to serious drawbacks that would affect the region and its economic growth especially from marine sector's contribution. Flooding, storm surge, sea rise would threaten coastal communities, infrastructures, settlements, and facilities when exacerbated. An economic crisis leading to disordered state of the society is imminent in these conditions. A study by the Asian Development Bank concerning four countries in the region, namely Indonesia, Philippines, Thailand, and Viet Nam as most vulnerable from GDP loss by 2010 (2.2%) annually, if without further mitigation or adaptation. Well above the world's 0.6%, *it could be dramatically worse, equivalent to 5.7% of GDP each year by 2100, if non-market impact (mainly related to health and ecosystem) was included and 6.7% if the chance of catastrophic events was also considered, way higher than the world's 2.2% and 2.6%, respectively* (ADB. April 2009).

The geographical features of Da Nang, a major city in Vietnam, makes it a vulnerable city to the impacts of climate change. Making the situation worse is the limited drainage capacity of its many rivers and streams. There are also many infrastructures that further contribute to the deterioration of its drainage capacity, affecting current, and water flow. Forest resources are depleted, thus reducing its natural function as buffers, protecting the city against strong winds and flooding. The high population density located in the urban districts also plays a role in its vulnerability. Da Nang, which is one of the most economically progressive areas in Vietnam, is being affected by serious damages and loss in terms of socioeconomic and environmental impacts caused by natural hazards such as typhoons and flooding that occurred in recent years. The latest data from the Scenario of Viet Nam Climate Change also shows increasing temperatures that will lead to sea level rise from 11.6 to 11.8 cm.

Recently, impacts of climate change are becoming increasingly evident as increasing weather disturbances and resulting natural disasters have been observed in the country. Based on available data and information gathered, the general trend of changes in climate such as temperature change, rainfall, drought, and frequency and strength of typhoon and salinity intrusion follow some of the general trend of changes in climate. These impacts are affecting agriculture, water resources with consequent impacts on food security, public health and safety as well as in natural resources directly affecting the communities.

Guimaras, which is an island province located in the middle part of the Philippines, is also prone to natural disasters such as storm surges, flood, landslides, and soil erosion due to the impacts of climate change. There are a number of environmental management issues that confront Guimaras prior to the ICM program. These include degradation of coastal resources due to various illegal activities such as cutting of mangroves, illegal fishing, destruction of coral reefs and sea grass beds, and extinction of endangered species; coastal erosion brought about by lack of preventive measures against sea level rise and sand mining activities; pollution from domestic sewage and solid waste; oil spills and operational discharges from sea vessels; and chemical contamination of rivers and coastal areas.

14.3 The Integrated Coastal Management (ICM) Framework

Strategic action programs in climate change adaptation and disaster risk reduction are vital now than ever as countries progress toward their respective sustainable development objectives. Through Integrated Coastal Management (ICM) development and implementation, countries in East Asia are progressing toward parallel Sustainable Development Goals (SDGs) especially in objectives related to poverty, food security, economic progress and ecosystems conservation.

The Sustainable Development Strategy for the Seas of East Asia (SDS-SEA) serves as a platform for cooperation and coordinated action among country and non-country partners and collaborators within the framework of Partnerships in Environmental Management for the Seas of East Asia (PEMSEA).

ICM is not just about environmental management. As adopted in global agreements such as the UN Conference on Environment and Development (UNCED) in 1992 and World Summit on Sustainable Development (WSSD) in 2002, ICM contributes to the protection of oceans and coastal areas and sustainable use of their living resources through its integrated and sustainable approach. Mending the divide brought about by the challenges of changing climate, ICM is a useful tool to address specific development challenges, through decision-making process that brings in a more integrated management approach. Through each stage in ICM cycle, a systematic process for each phase from planning to initiating, developing to implementing, monitoring and evaluating strategies, programs, and services is put in place to respond to the challenges faced by coastal communities.

According to Chua (2006), ICM's purpose is to increase efficiency and effectiveness of coastal governance in terms of ability to achieve the sustainable use of coastal resources and of the services generated by the ecosystems in coastal areas. It aims to do this by protecting the functional integrity of these natural resource systems while allowing economic development to proceed.

What sets ICM apart from other coastal management framework are those principles such as adaptive management, integration and coordination and ecosystem-based management approaches. Through an ecosystem-based management approach, ICM considers human activities in its management framework and their impacts on the ecosystem, rather than merely a management of components of ecosystem. Thus, it has become a very useful guide to build on the ground capacity of national and local governments in promoting initiatives and programs through policy, strategies, and action plans that are consistent with sustainable development principles. Current ICM practices reflect broad-based approaches and operational strategies that encourage effective instruments and best practices in governance frameworks.

As seen in Fig. 14.1, the SDCA framework consists governance as well as several issue-specific management systems critical to achieving the SDGs, including climate change adaptation and disaster risk reduction. It includes components on policy reforms, strategies and action plans, institutional mechanisms such as interagency

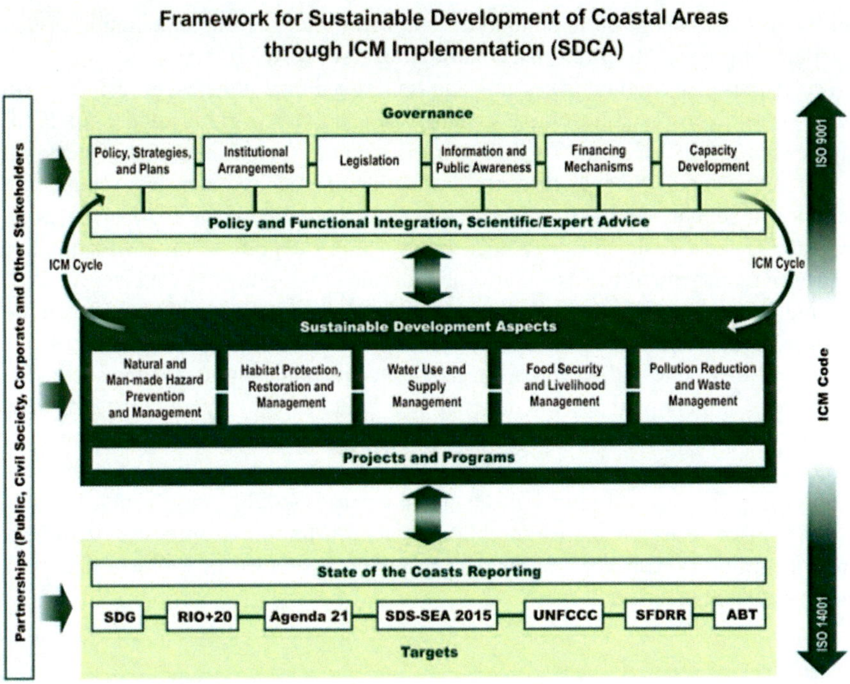

Fig. 14.1 Sustainable Development of Coastal Areas (SDCA) Framework, PEMSEA

and multisectoral arrangements, development and implementation of national legislations and local administrative orders, public awareness, information management, sustainable investment measures and financing, and capacity building. These governance mechanisms are coupled with on-the-ground actions that integrate policies into operating management systems. While it is deemed necessary to have coordination with the national authority, ICM also intends to build the preparedness of the local government and its stakeholders to respond to both man-made and natural disasters. The ICM approach facilitates the identification early interventions and timely response both at the operational and strategic levels.

14.4 ICM in Da Nang, Vietnam

ICM program in Da Nang started in 2000 and since then has contributed to adaptation and mitigation measures in addressing the challenges to responding to the impacts of climate change in Da Nang. These include limited planning and technical skills of local staff, limited knowledge and awareness on the integration of natural disaster mitigation and management into social development, lack of scientific basis for integration of climate change adaptation into the socioeconomic development plan, and limited assessment on livelihood development after the implementation of the resettlement program for high risk communities (ACCCRN 2009).

Da Nang is instrumental in crafting and issuing national policies, guidelines, and action plans in response to climate change as a proven working model for ICM implementation in Vietnam. These documents relate to environmental protection, energy savings and efficiency measures, reduction in greenhouse gas emissions, and activities to support green growth. One of the key outputs of the ICM program in Da Nang is the Coastal Strategy, that serves as an integrative framework for the long-term management of coastal area and marine environment in Da Nang.

Guided by the principles of sustainable development and application of integrated planning and management approach, the response of Da Nang to climate change is well in line with the broad-based multistakeholder approach and socioeconomic development plan and the national government's strategy on climate change and disaster risk reduction. The ICM program has contributed to the implementation of adaptation and mitigation measures to alleviate and respond to the impacts through improving institutional mechanisms such as establishing a coordination mechanism for disaster response, integrating scientific findings from research studies, strengthening local capacity and increasing public awareness as well as promoting energy-efficient practices.

14.5 ICM in Guimaras, Philippines

In August 11, 2006, an oil spill incident occurred in Guimaras. This unfortunate incident turned into an opportunity to open Guimaras to the international community. Since then Guimaras started ICM and adopted the SDCA framework and since has been the main vehicle to implement various programs and projects in coastal areas for sustainable use of resources and reducing conflict. Due to limited human and financial resources, there was a lack of comprehensive plan on Disaster Risk Reduction and Management in the province. Through technical support and financial support from various projects such as the Yeosu project in partnership with the provincial government, a Provincial Disaster Risk Reduction and Management Framework has been developed, which subsequently became the Guimaras DRRM Plan guided by the principles based on the National DRRM Framework, Sustainable Development, and ICM. The management strategies of the framework respond to specific hazards in the province such as storm surges, flood, landslide, erosion, and oil spill.

As the results of vulnerability assessment revealed the vulnerable areas for disasters, four priority projects were identified as actions to respond to disaster preparedness: (1) Building Drainage System in Flood-prone Areas, (2) Training on Emergency Operation for DRRM Staff, (3) Establishing Early Warning System, and (4) Ecosystem Protection and Conservation of Critical Habitat. Due to the changing climate and environmental degradation, the key critical habitats-coral reef, sea grass bed and mangrove in the province need to be conserved and protected as part of disaster prevention efforts. Other recommendations include the establishment and implementation of effective management of Marine Protected Areas along the eastern coast of Guimaras Province, thus the focus of the Second and Third phase of the Yeosu project.

As ICM cuts across various multiple issues through implementation of programs, projects, and activities, the scope of work involves development of comprehensive conservation and management plans for key critical habitats via mapping of three proposed MPAs in Tumalintinan Point MPA and Nadulao Island Gakit-gakit Shoal in San Lorenzo and Pamangkulan Fish Sanctuary in Sibunag and delineation of management boundaries as well as conduct of survey of people's awareness. Through conservation of critical habitats which act as natural coastal defense from storm surge and tsunami as well as fisheries nursing grounds, the protection of these critical habitats provides Guimaras with coastal protection, enhancing fisheries and increased eco-tourism through charismatic marine mammal conservation and protection.

14.6 Lessons Learned

There were several lessons learned from these two case studies regarding the local government's implementation of climate change adaptation and disaster risk reduction.

First, institutional arrangements operationalizing interagency and multisectoral coordinating mechanisms can effectively promote joint planning and implementation if under city leadership that can facilitate cooperation between and among line agencies and levels of governments. Stakeholder participation is also vital in planning, implementing, evaluating, and continually improving programs as well as in generating and mobilizing human and financial resources in the development and implementation of action plans and programs on climate change.

Second, to effectively reduce loss and damage in lives, property, livelihood, and local economy, there should be adequate capacity and responses that are efficiently implemented in a timely manner. Local governments and local communities are usually the first that can be affected by the impacts of climate change, but they can also serve as the first responder on the ground. Building their capacity and public awareness through trainings, workshops, and public information could increase not only the level of consciousness about the risks but also their ability to act properly to protect themselves and reduce exposure or loss. Translating findings from scientific studies into policymaking and decision-making process can also enhance the capacity of the government to effectively respond to climate change.

Lastly, ICM program has served as a broad planning and management framework for sustainable development and laid the foundation for other national and international partners to work together and have confidence in the climate change projects and programs of Danang and Guimaras.

14.7 Conclusion

Adhering to the framework of SDCA, the implementation of various ICM programs in Danang and Guimaras has addressed various Sustainable Development Aspects on Habitat Protection Restoration and Management, Food Security and Livelihood Management, Pollution Reduction and Waste Management, and the Natural- and Human-Induced Hazard Prevention and Management. In the case of Guimaras, the efforts have resulted to improved critical habitats and fish stock and to the socioeconomic development plan, and limited assessment on livelihood development in the case of Danang.

Evident in both cases is that the scope of ICM implementation is growing while addressing various aspects of sustainable development of the coasts and marine areas. Through ICM programs of the local governments, a balance is maintained between economic development an environmental protection of both sites. Though various interventions and adaptation have responded to climate change impacts, there is still a need to strengthen the multiagency and multisectoral nature of coordinating mechanisms and the establishment of regular mechanisms for monitoring.

References

ACCCRN (Asian Cities Climate Change Resilience Network) (2009) In: Hazard, capacity and vulnerability assessment in Da Nang

Asian Development Bank (2009) In: The economics of climate change in southeast asia: a regional review

Chua T-E (2006) In: The dynamics of integrated coastal management: practical applications in the sustainable coastal development in East Asia. Global Environment Facility/United Nations Development Programme/International Maritime Organization Regional Programme on Building Partnerships in Environmental Management for the Seas of East Asia (PEMSEA), Quezon City, Philippines, p 468

Chapter 15
Landslide Disaster Management in Sri Lanka

Asiri Karunawardena

Abstract Sri Lanka experiences landslides and slope failures during the periods of heavy rainfall, and their frequency and intensity have been observed to be increasing because of climate change impacts. This paper describes the role of National Building Research Organization (NBRO) that plays as the national focal point for landslide risk management. NBRO started landslide investigations since 1985. Identification of landslides, susceptibility assessment and mapping were the pioneering works of NBRO and implementation of Landslide Hazard Zonation Mapping Project (LHMP) resulted in a valuable outcome of preparing and updating hazard zonation maps over the past 25 years. Awareness creation, mitigation, monitoring and early warning activities in the country are other important functions of NBRO. Monitoring networks have been established for early warning. As a proactive action to minimize the adverse impacts of landslides, structural mitigation is being carried out by NBRO and several technically advanced slope stabilization techniques have been introduced with the international assistance as discussed by this paper giving some salient details.

Keywords Landslide management, landslide disaster mapping · Early warning · Mitigation

15.1 Introduction

Landslides in hilly or mountainous terrains are a common natural phenomenon in many parts of the world. A landslide event is defined as "the movement of a mass of rock, debris, or earth (soil) down a slope (under the influence of gravity)". The word "landslide" also refers to the geomorphic features that result from the event. During the past few decades, landslides occurred with an increasing frequency in the hill country of Sri Lanka. Probability for landslides in the hill country, underlain by highly folded, fractured and weathered metamorphic rocks, is high. Hilly areas prone to landslides cover an approximate extent of 10,000 sq.km, which is about 20% of the

A. Karunawardena (✉)
National Building Research Organisation, Colombo, Sri Lanka
e-mail: asiri13@hotmail.com

H. Tatano and A. Collins (eds.), *Proceedings of the 3rd Global Summit of Research Institutes for Disaster Risk Reduction*, Disaster and Risk Research: GADRI Book Series,
https://doi.org/10.1007/978-981-15-8662-0_15

total area of the island and occupied by about 30% of the total population of Sri Lanka, as shown by Fig. 15.1. These hilly areas lie within ten administrative districts, namely Badulla, Nuwara Eliya, Kegalle, Ratnapura, Kandy, Matale, Kaluthara, Mathara, Galle and Hambantota.

Landslide disasters cause loss of lives and severe damage to property and the environment, and their impact to economy and the society is significant. The loss of lives between year 2000 and 2017 is shown by Fig. 15.2. After the landslides in 1984, the Government of Sri Lanka assigned National Building Research Organisation (NBRO) to study landslides in the country. Since then, NBRO has taken many initiatives in landslide risk reduction and presently NBRO holds the mandate to manage

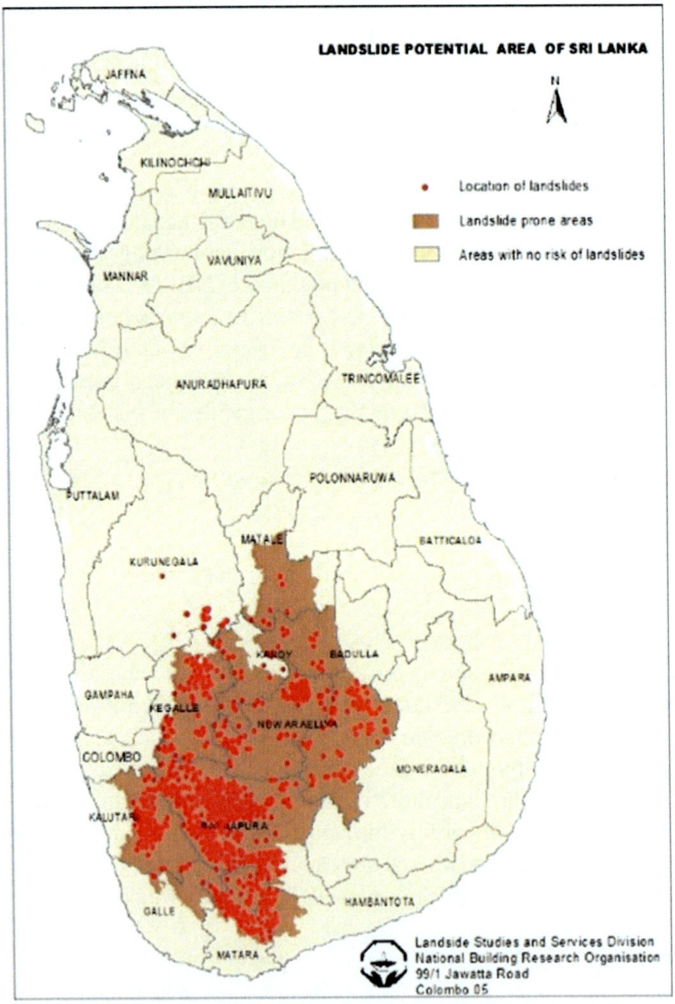

Fig. 15.1 Landslide distribution map

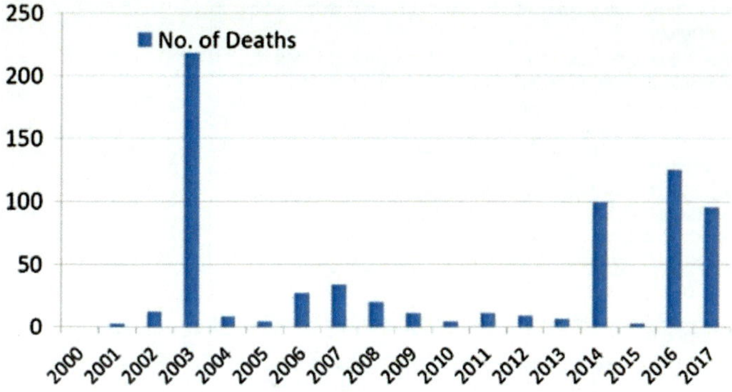

Fig. 15.2 Number of lives lost due to landslides (2000–2017)

the landslide disaster risk in the country. NBRO conducts landslide risk assessment, mapping, monitoring, early warning and mitigation activities and contributes to related policy planning and development activities as an important stakeholder.

15.2 Landslide Investigations and (LHMP)

NBRO launched Landslide Hazard Zonation Mapping Project (LHMP) in 1986 in view of establishing sustainable long-term and short-term mechanisms for landslide hazard management in Sri Lanka. In this project, landslides are continually investigated and then their susceptibility levels are demarcated as zones in maps. By now, the project has covered most landslide-prone areas in the country.

The project uses Landslide Hazard Potential Evaluation Methodology, which is a probabilistic approach based on the physical characteristics and the professional judgement. The factors Bed Rock Geology and Geological Structures, Slope Range and Category, Hydrology and Drainage, Land Use and Management, Surface Deposits and Landform are considered as the major causative factors of Sri Lankan landslides (User Manual—Landslide Hazard Mapping in Sri Lanka 1995). Since the impact of causative factors on landslide potential varies, each of the above factors is rated by weightages according to the statistical analysis and expert knowledge. The approach used in assigning the weightages is presented by Fig. 15.3.

The digitized coverages of geological structure, hydrology, land use, land form, surface deposits and slope were integrated to create an inferred map of landslide potential. For each polygon of this inferred map, the numerical hazard ratings for each factor were added to compute the overall hazard rating. The polygons of this inferred map are dissolved into different hazard zones using the criteria given in Table 15.1. These criteria were arrived at by trial and error after spatially matching data of numerous past landslides and inferred instability (User Manual—Landslide Hazard Mapping in Sri Lanka 1995). The dissolved map of inferred landslide potential

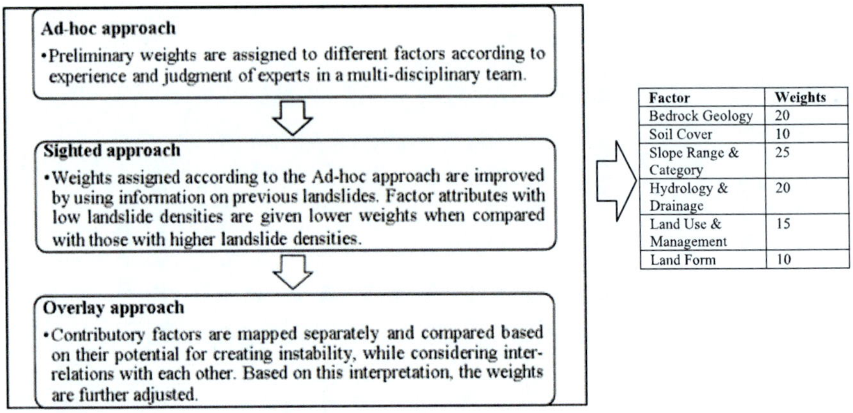

Fig. 15.3 Approach used to for assigning weights

Table 15.1 Different hazard zones

Overall hazard rating	Hazard zone	Description
R< =40%	1	Safe slopes
40<R< =55	2	Landslides not likely to occur
55<R< =70	3	Modest level of landslide hazard
70<R	4	Landslides are expected

is overlaid by the coverages, inaccessible slopes, past landslides and rivers. The coverages, small streams, major and secondary roads, railways and locations were also added to input as much as information possible.

Maps to the scales of 1: 50,000 and 1:10,000 are prepared and continually updated. These maps serve NBRO in their preparedness, early warning and mitigation work. The maps are freely available in the web and frequently referred in land use planning and in the approval of construction and development projects. The areas covered by the LHMP are shown in Fig. 15.4 and a sample of district-level landslide hazard zonation maps is shown in Fig. 15.5.

15.3 Landslide Early Warning and Awareness Creation

NBRO uses computer simulation to forecast the possibility of landslides and a network of automated rain gauges was established in vulnerable areas to acquire real-time rainfall data. This network is being expanded to cover other vulnerable

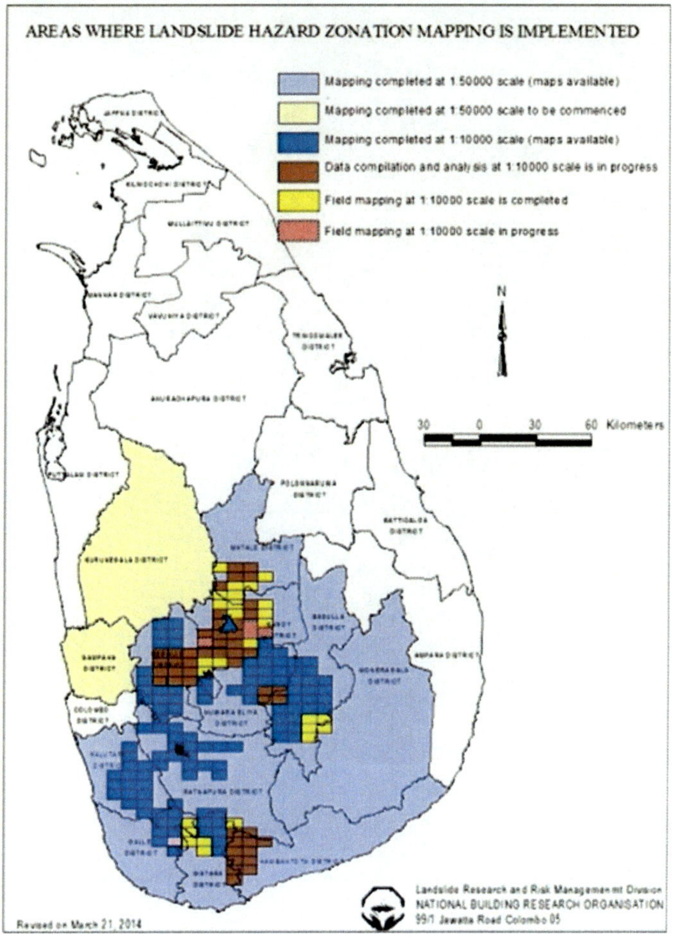

Fig. 15.4 Area where landslide hazard zonation mapping is implemented

areas as well. Manual rain gauges have been given to volunteers in some vulnerable areas for monitoring rainfall and inform NBRO at times of excessive rainfall. NBRO is in constant contact with Department of Meteorology and other institutions gathering rainfall data. All the gathered rainfall data are analysed on a continuous basis and when rainfall intensities reach to certain threshold values early warnings are issued (Refer Table 15.2). Incidents are now in record how such warnings saved many lives. There are communities living in remote villages where communication in emergencies is difficult, and NBRO is establishing community-based early warning systems to make them self-reliant. In order to fortify early warning systems, NBRO has already ventured into ground movement monitoring at vulnerable locations like banks of highways and slow-moving landslides.

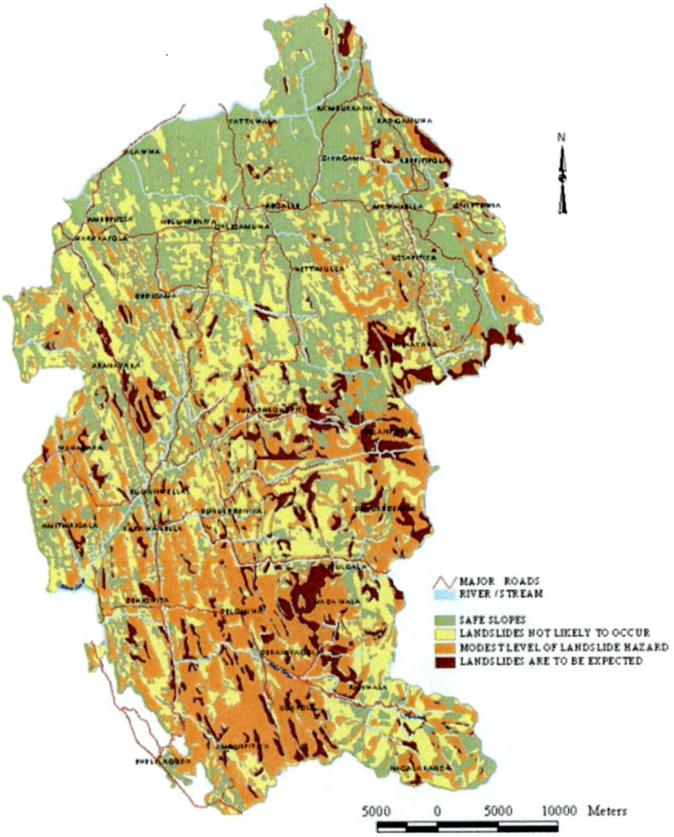

Fig. 15.5 Landslide hazard zonation map of Kegalle district (A sample of district level hazard map)

Table 15.2 Threshold values

Warning levels	Colour code	Description of warning
75 mm	**Yellow**	Be on alert on the possibility of landslides, rock falls, subsidence and cut slope failure if the rainfall exceeds 75 mm and prevailing bad weather conditions continue within next 24 h
100 mm	**Amber**	If the rainfall exceeds 100 mm and continues within next 24 h, danger of landslides and cut slope failures exists
150 mm	**Red**	If the rainfall exceeds 150 mm and continues within next 24 h or if the rainfall exceeds 75 mm within an hour, evacuate to a safe place

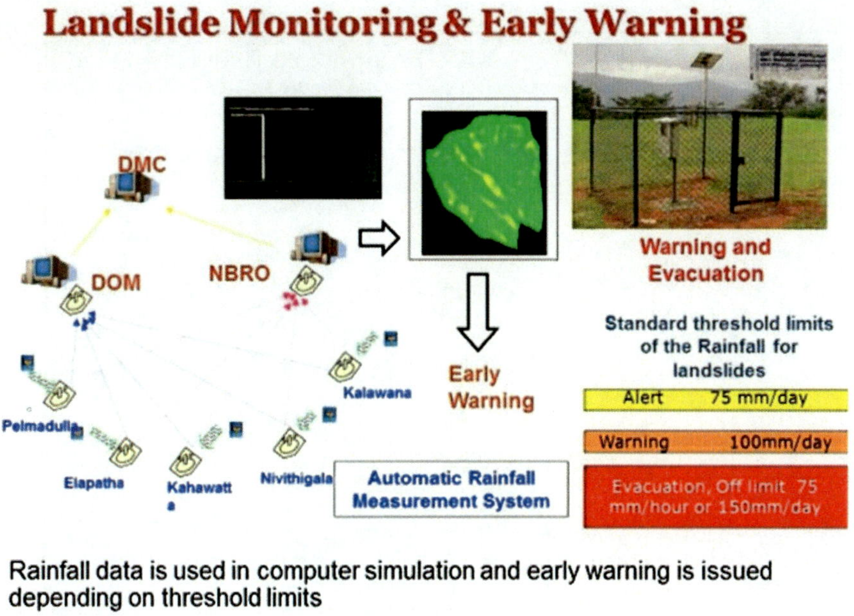

Fig. 15.6 Landslide early warning system

In hazardous situations, early warning is issued according to set rainfall threshold limits through the 24/7 emergency operations center of the Disaster Management Centre where the dissemination of early warning is carried out through the media and dedicated communication systems to both endangered communities and concerned authorities for evacuation of vulnerable persons to safety (Refer Figs. 15.6 and 15.7).

School children, local authorities and communities in general are made aware of landslide hazard in respective areas they live by conducting awareness programmes and mock drills where they learn about the causes that trigger landslides, signs that indicate landslide movement, methods to prevent landslide occurrence and importantly, how to evacuate swiftly and systematically in an emergency. NBRO made many publications and also some games for school going children for effective dissemination of landslide hazard risk.

15.4 Approval of Construction and Development Projects

Due to non-engineered construction, natural slopes are disturbed and result in landslides and slope failures (Refer Fig. 15.8). In order to prevent non-engineered construction, NBRO provides guidelines for land use planning, construction and other development activities (Refer Fig. 15.9). At present, the local government authorities issue a building permit or a construction activity or project approving

Fig. 15.7 Locations of automated rain gauges

Fig. 15.8 A non-engineered housing construction

Fig. 15.9 An engineered housing construction

agencies grant approval for a development project in an area susceptible to landslides only after the concurrence of NBRO has been obtained by way of a Landslide Risk Assessment Report and recommendations therein. NBRO district offices in ten landslide-prone districts accept applications, investigate and issue these reports. This process has effectively reduced haphazard construction and development in vulnerable areas, curtailing unplanned and non-engineered construction to a great extent, and thereby, lowering man-made hazards.

15.5 Structural Mitigation

The causes of landslides are usually related to instabilities in slopes and it can include geological factors, morphological factors, physical factors and factors associated

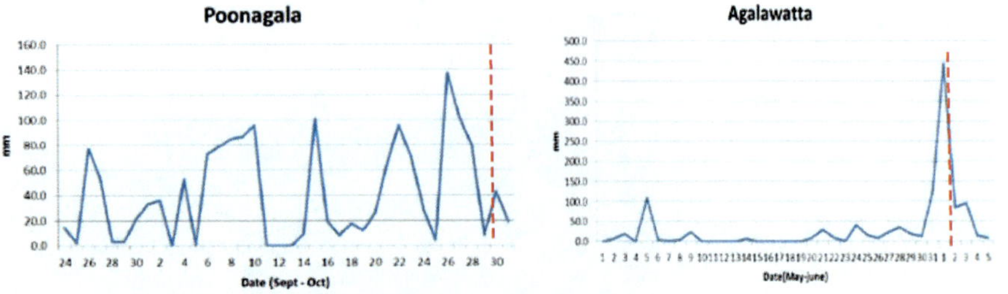

Fig. 15.10 Recorded rainfall before the landslide at Meeriabedda (2014-10-29) and Agalawatte in Kalutara District (2014-06-01)

with human activity. In Sri Lanka, intense rainfall may be regarded as the main triggering factor of landslides because most landslide disasters in the country occurred in monsoon period. Figure 15.10 shows the recorded rain fall experienced before the landslides in Meeriabedda and Kalutara area. With a clear understanding of the causes and mechanics, landslides can be physically stabilized or controlled by the application of structural mitigation methods.

Structural mitigation methods can be broadly categorized into three types: geometric methods, in which the geometry of the hillside is changed (in general the slope); hydrogeological methods, in which an attempt is made to lower the groundwater level or to reduce the water content of the material; chemical and mechanical methods, in which attempts are made to increase the shear strength of the unstable mass or to introduce active external forces. All of these slope stabilization methods have many alternatives to their shape, structure and covers. The selection of the most appropriate structural mitigation measures to be adopted in specific situations is based on the detailed geotechnical design using geotechnical/geological investigation and monitoring results. The procedure that we are adopted in landslide mitigation is presented in Fig. 15.11.

Initially detailed geotechnical investigation and instrumentation are carried out. Sophisticated technology is used to collect more realistic and reliable data at both investigation and monitoring. Different techniques are used to obtain topographical data. Investigation data and core samples are carefully observed to identify any indications of failure surface. After the completion of the detailed investigations, automated monitoring instruments are installed at the site for further verifications of test data (e.g. depth of slip surface/s) and to gather further data for the design of countermeasures (e.g. groundwater behaviour of the site alongside rainfall and rate of landslide movement). The automated monitoring systems are used to establish relationship between the rainfall and groundwater level and movements of landslide with which are essential information for design work.

Based on the results obtained from detailed investigation and descriptive idea formulated from different landslide surveys, landslide mechanism is identified as accurate as possible so that stability analyses are carried out to determine the scale and quantity of landslide countermeasures required to maintain the stability of the landslide.

Fig. 15.11 Landslide mitigation procedure

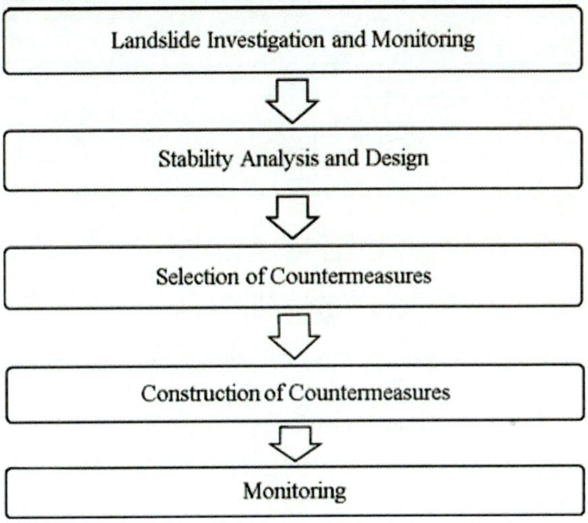

Selection of the suitable countermeasure work depends on the technical effectiveness, financial viability, environmental impact and amount of maintenance required. A wide range of countermeasures such as change of geometry, surface drainage, subsurface drainage, counterweight embankments, piling and ground anchors is considered.

Unlike conventional practices, several combinations of countermeasures are considered based on the stability analysis. Each is examined to decide a couple or more appropriate combinations of countermeasures in terms of both economic efficiency and design requirements. From the more appropriate combinations of counter measures, an optimum combination is selected through careful examination and comparison. Figure 15.12 shows the procedure followed to select the appropriate countermeasure by repetitive stability analyses.

The detailed design carried out according to above procedure for Badulusirigama landslide near Uva Wellassa University is shown in Fig. 15.13.

15.5.1 *Large-Scale Landslide Mitigation Work Carried Out by NBRO*

The first large-scale landslide structural mitigation work carried out in Sri Lanka to stabilize unstable slope is the Watawala landslide. Beragala and Pussellawa are other examples of structural mitigation work carried out during the period from 1990 to 2000. Then, many landslides occurred due to high rainfalls experienced in the country in 2002, 2003 and 2007 period. As a result of that, the need for stabilization of unstable slopes became very apparent, and therefore, NBRO stepped into mitigation of high-risk major landslide and rock fall sites and stabilization of identified potential

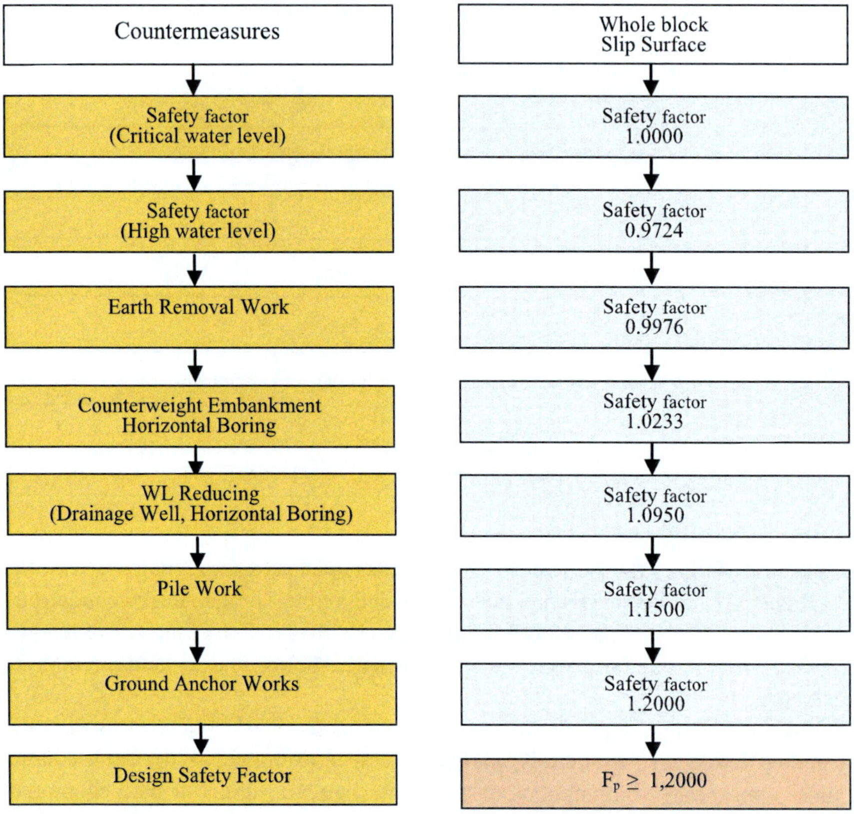

Fig. 15.12 Procedure used for selection the appropriate countermeasure

unstable slopes in Nuwara Eliya, Kandy, Matale and Badulla districts. In all the above sites resettlement, option was found to be not an economically viable option of risk removal. On the contrary, option of structural landslide mitigation was found to be more cost effective. The paper presents the details of some of major landslide mitigation work carried out by NBRO.

15.5.1.1 Stabilization of Watawala Landslide

NBRO pioneered structural mitigation of landslides in Sri Lanka by providing technical inputs to stabilize the Watawala landslide. Watawala landslide occurred in June 1992 damaging nearly 100 m length of the Colombo–Badulla railway track between Galboda and Watawala as shown by Fig. 15.10. The landslide was found to cover an area of approximately 22,920 m^2 and about 322,700 m^3 slide volume. The investigation and design clearly showed that remedy lies in very effectively draining the slope so as to lower the piezometric profile. Therefore, the stabilization of mitigation

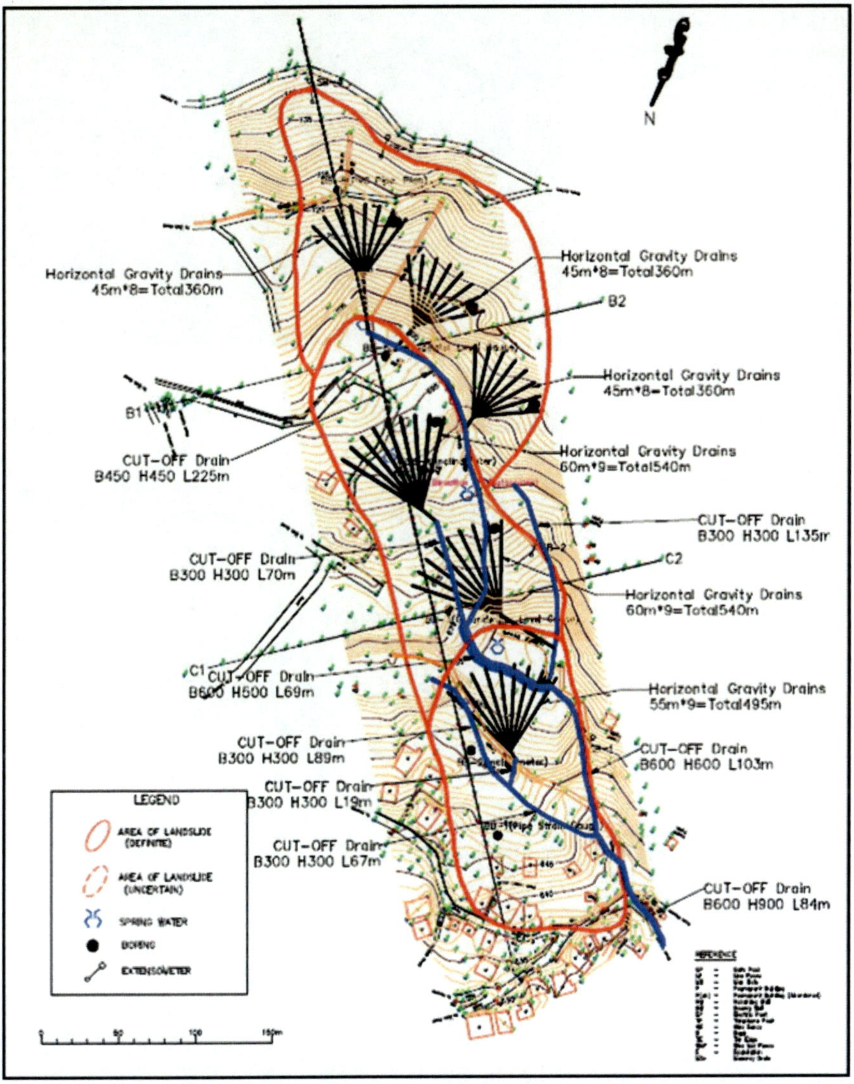

Fig. 15.13 Design of mitigation measures at Badulusirigama landslide

of Watawala landslide was carried out by constructing well-designed subsurface and surface drainage system (Figs. 15.14 and 15.15).

Fig. 15.14 Damage to the railway track

Fig. 15.15 Present situation at Watawala after mitigation

15.5.1.2 Stabilization of Peradeniya Landslide

Peradeniya landslide is located at Peradeniya town, by the Colombo–Kandy main road. This landslide occurred in November 2006 destroying a few boutiques in the town and interrupting road traffic for weeks. Peradeniya is the gateway to the historical city of Kandy. Since the year 2009, the Government of Sri Lanka had allocated required funds to mitigate this landslide for the benefit of the users of Colombo-Kandy main road and the dwellers and commuters of the Peradeniya town.

The mitigation work was started in December 2008 by NBRO. Based on investigation data, stability analysis was carried out to select the appropriate structural counter measures to improve the stability and some results of one of the critical sections are shown in Figs. 15.16 and 15.17. The applied structural mitigation measures for mitigating this landslide include reshaping the slope, removing unstable rock by blasting, rock bolting, construction of surface drainage, erosion control slope protection measures and retaining walls. Activities of the phase 1 of the project were completed in 2012. The appearance of site before and after the structural mitigation is shown in Figs. 15.18 and 15.19.

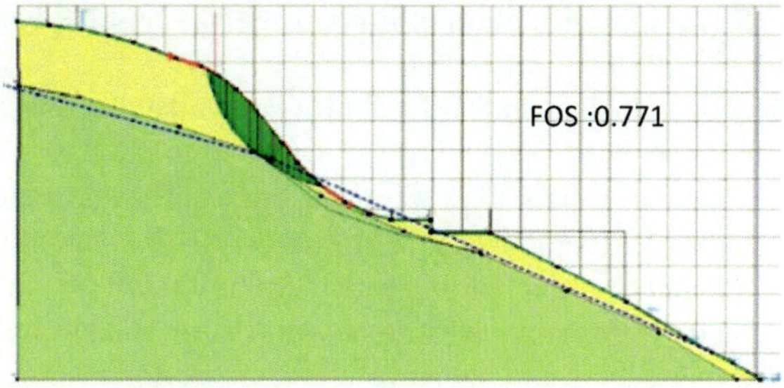

Fig. 15.16 Results of the stability analysis for existing slope under critical condition

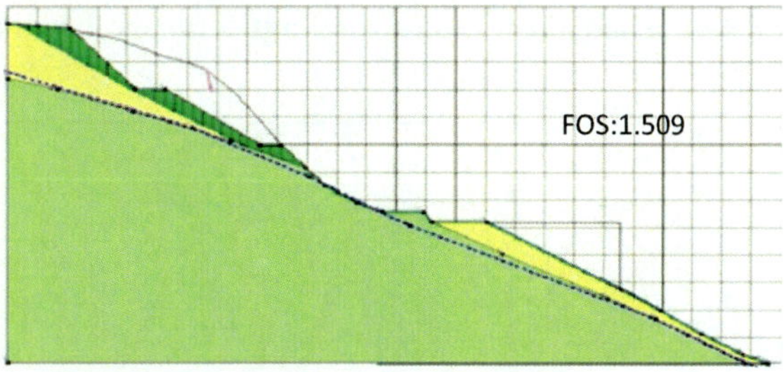

Fig. 15.17 Results of stability analysis with proposed mitigation methods

Fig. 15.18 Peradeniya landslide

Fig. 15.19 Peradeniya landslide after mitigation

15.5.1.3 Stabilization of Padiyapellella Landslide

Padiyapalella landslide is located at Padiyapalella town, by the Walapane–Hanguranketha main road. This landslide activated in 2008 resulting of evacuation of injured persons and others to safety, and proposals were made to mitigate the hazard to make the Padiyapallale township safe. NBRO started mitigation work in 2009 and completed to the relief and satisfaction of town dwellers in 2012. Many mitigation methods were deployed in this work, namely land reshaping, retaining structures, turfs, surface drains, subsurface horizontal drains and soil nailing. The plan view of the designed structural mitigation method is shown in Fig. 15.20. The appearance of site before and after the structural mitigation is shown in Figs. 15.21 and 15.22.

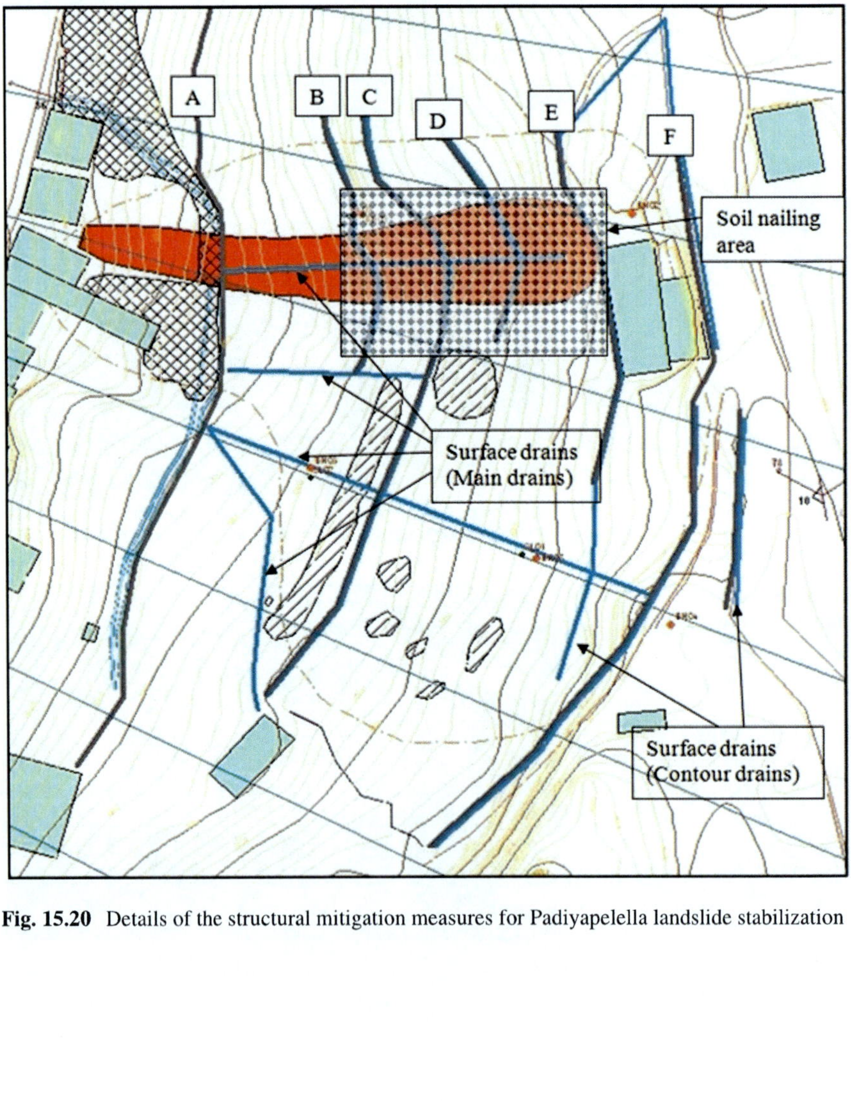

Fig. 15.20 Details of the structural mitigation measures for Padiyapelella landslide stabilization

Fig. 15.21 Padiyapallale landslide

Fig. 15.22 Padiyapallale landslide after mitigation

15.5.1.4 Stabilization of Garandiella

Garandiella landslide is located by the 41st km post on Nuwara Eliya–Gampola main road in Kothmale Division. Mitigation of this landslide was commenced on August 2011 for reducing the landslide risk on the users of the Nuwara Eliya–Gampola main road, Kothmale reservoir, 230 families living in the Ramboda Grama Seva Wasama in Kothmale Division and the built environment in the Kothmale town. Investigation reveals that the unstable slope areas consist of thick overburden and the ground water rise to almost up to the surface level in the rainy period. The results of the stability analysis indicate that both upper and lower part of the slope will become unstable under the critical condition. The results of stability analysis of the upper slope under the assumed critical condition are shown in Fig. 15.23. The results of stability analysis for the same slope with proposed mitigation measures are shown in Fig. 15.24.

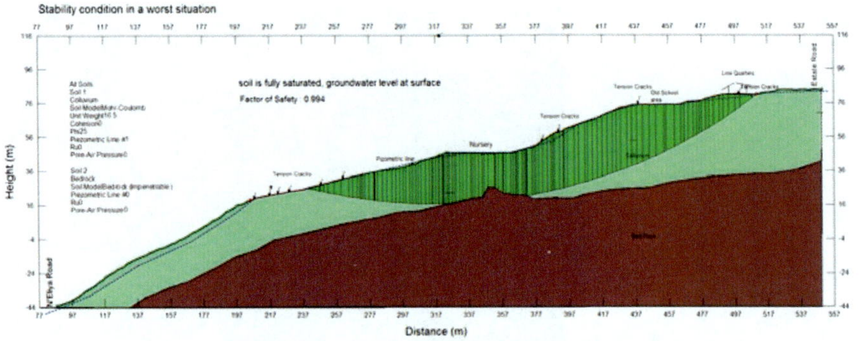

Fig. 15.23 Results of the stability analysis for upper slope for assumed critical condition

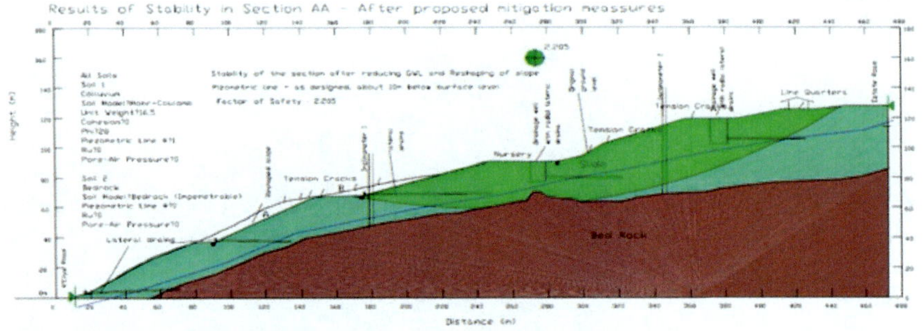

Fig. 15.24 Results of the stability analysis with proposed mitigation measures

15.5.2 Landslide Mitigation Projects in Progress

NBRO has assessed severity of landslides in the country and developed plans to mitigate them in a prioritized manner creating resilience to landslides. Mitigation of deep seated landslide in Punchi Raththota in Matale district, stabilization of unstable slopes near the Diyatalawa main bus stand and slope stabilization in Passenwatta, a residential area in Gampaha, are projects nearing completion. Landslide Disaster Protection Project of the National Road Network with the support of Japan will be implemented where the Japanese technology will be applied to mitigate landslide risk along the major roads in central highland in the country. Also, four major landslides in the districts of Kandy, Matale, NuwaraEliya and Badulla will be mitigated under Japanese Technical Cooperation Project.

Under the Climate Resilience Improvement Programme of the World Bank, stabilization of unstable slopes in and around 18 identified schools in Kandy District and stabilization of unstable slopes in major roads network are in progress. Figures 15.25, 15.26 and 15.27 show the details of the mitigation work carried out at the failed slope at Southern Express Way.

Fig. 15.25 Failed slope at southern expressway

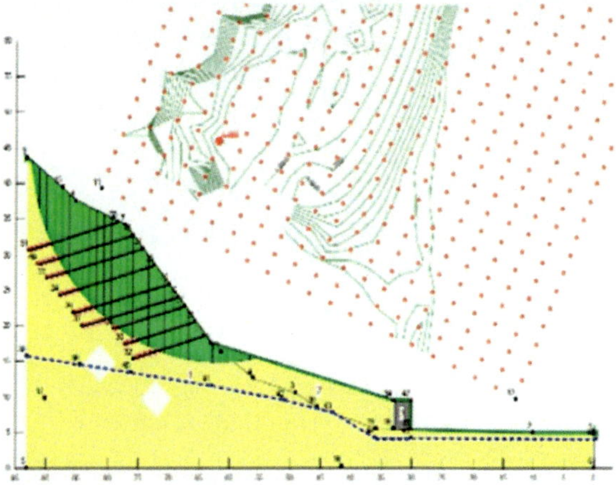

Fig. 15.26 Design of mitigation measures at failed slope at southern expressway

Fig. 15.27 Slope at southern expressway after mitigation

15.6 Conclusion

NBRO played the main role to lower adverse impacts of landslide disasters in Sri Lanka. It is appreciable that many international organizations are coming forward with assistance and JICA and World Bank assistance have been very significant. NBRO should pursue learning advanced and more effective mitigation technologies in view of achieving resilience countrywide, making Sri Lanka safer.

References

User Manual - Landslide Hazard Mapping in Sri Lanka Landslide Hazard Mapping Project (1995)

Chapter 16
Framework to Promote Multi-disciplinary Culture of Landslide Studies and Management

A. A. Virajh Dias, N. N. Katuwala, and P. V. I. P. Perera

Summary A multi-aspects and policy framework are a tool with numerous variations and contexts which are used to make conceptual distinctions and organize ideas toward the focusing subjects. One way, it is a simple tool for practicing approach. Other way it looks as a supportive arm of establishing various important strategies to practice in a society. The most important fact is that the validity of these frameworks has not been tested through interaction with the community and the policy makers. Secondly, these methods generally have their uses, but are unable to be implemented if the relative probability of the particular risk appears to be too low to be considered. Other reasons are usually combinations of above which facilitate sustainable development under disaster risk. Framework to promote multi-disciplinary culture of mitigation of landslides needs various understandings of scientific proven approach on investigation, planning, and execution work. Understanding of overall needs and requirements of a community at risk due to landslides broadly explains sustainability, social adoptability, and uniqueness. Mountain development policies ensure that population does not exceed carrying capacities of an area and will serve with low probabilities of being affected by landslides. Community-centered monitoring and early warning systems are highly adaptive mechanisms for a community living with landslides proven zones, so that the local people and village officers are capable of disaster preparedness and also operating the system in an emergency. Therefore, multi-disciplinary culture of the landslide disaster studies and management is giving an opportunity to facilitate various technical and social considerations erected in a common framework which addresses planned resilient human habitat in areas that are prone to landslides.

World Center of Excellence on Landslide Disaster Reduction (Model Policy Framework Standards and Guidelines), International Program of Landslides (IPL-ICL), 2014–2020.

A. A. Virajh Dias (✉)
Natural Resources Management and Laboratory Services (NRM & LS), Central Engineering Consultancy Bureau (CECB), Colombo, Sri Lanka
e-mail: aavirajhd@yahoo.com

N. N. Katuwala · P. V. I. P. Perera
Environmental Studies (NRM & LS, CECB), Colombo, Sri Lanka
e-mail: nkatuwala@gmail.com

H. Tatano and A. Collins (eds.), *Proceedings of the 3rd Global Summit of Research Institutes for Disaster Risk Reduction*, Disaster and Risk Research: GADRI Book Series,
https://doi.org/10.1007/978-981-15-8662-0_16

Fig. 16.1 Sustainable geography of community living

Keywords Policy framework · Sustainability · Landslides · Disaster management

16.1 Introduction

Landslides are frequently occurring disasters during heavy rains in various parts of the world. To a layman exposed to a landslide, it could be a rapid event but for those who are specialized in landslides, such a statement would be a mere excuse. Indeed, it is someone's business to investigate, instrument, monitor, and analyze the slopes so as to capture the signals of instability well in advance. We, however, mostly deal with events in stable slopes gifted to us by nature which were earlier considered safe zones on our judgments are getting transformed into danger zones because of human violence against our mountain systems. It is important to underscore this point in the present day context because most people including media often unfairly blame society for landslides occurrences. Before moving in to understanding the importance of multi-disciplinary culture of the landslides, it is wise to study the unique nature and patterns of de-stability begging in nature. The scientific concept of stable and natural slope segments is defined as Fig. 16.1.

16.2 Scientist Interactions of Slopes

Geoscientists are the foremost group who understand the mechanisms of landslide behavior and the process of degradation of natural slope. They continuously do research to study the mechanisms, behavior, stability, and the potential hazards in various continents of the world with the use of advanced and multi-spectral approaches of sciences. Use of geographical information systems and remote sensing technique largely contribute for landslide disaster prone zone identification, prediction, and management. Although, large number of unforeseen factors of nature will indicate the ground deformities and in-situ stress, but are limited to foresee factors

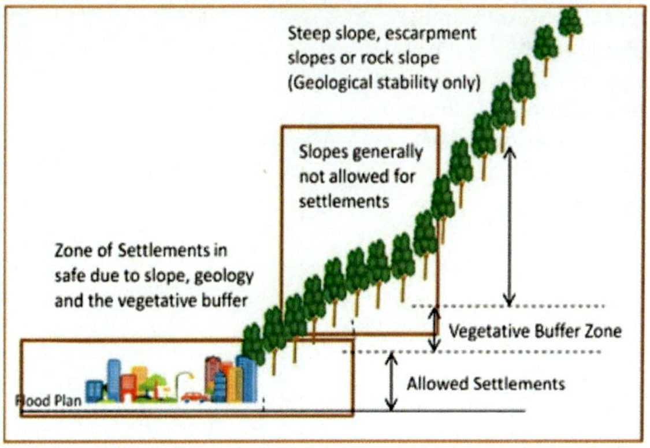

Fig. 16.2 Stability of hill country slopes. No recharge or artificial water infiltration due to lack of infrastructure development; more safe zones of slopes

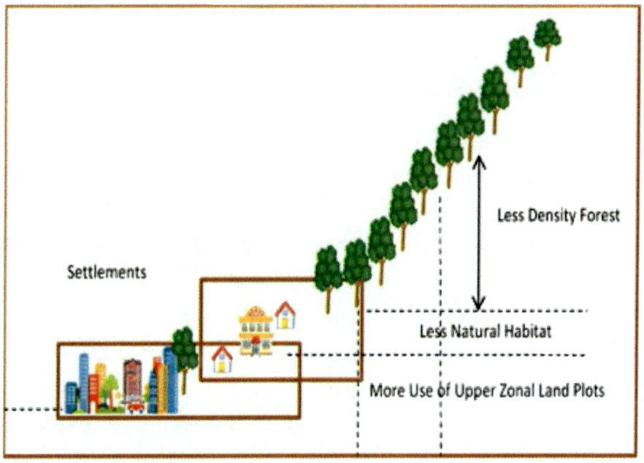

Fig. 16.3 The stage approach of moving upward toward the slope; initiated preliminary due to land encroachers; or lack of strategic approaches on new settlement planning in hill slope management against landslide risk

responsible for ground instability. Understanding the nature and stability are primary concerns in community living in hill slopes as shown in Figs. 16.2 and 16.3.

Number of intrinsic and extrinsic factors (Hutchinson 1988) are responsible for ground instability including unplanned or non-engineered construction or aggressive human interventions on slopes. However, geoscientists are capable of recognizing and demarcating causes and effects of ground for numerous occurrences of landslides and record the same according to scientific discipline (Rupasinghe et al. 2014). Change in slope or geomorphology (from naturally or man-made) may causes disequilibrium between earth mass and transporting material (flow debris and sediments). Therefore, various classifications of landslide are made considering physical phenomena and evidences of nature. There are no clear evidences of quantification of probability of reoccurrence of landslides due to various unforeseen factors inherent to natural slope. It is noted that large number of landslides prone areas are initiated due to human interventions on natural slope or inability to work with nature's stability as shown in Fig. 16.4.

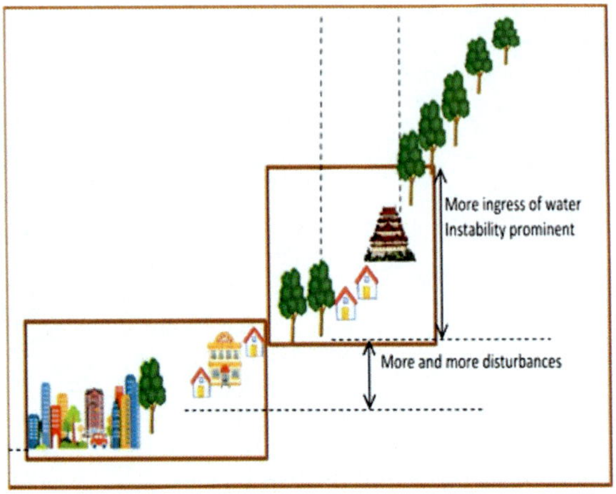

Fig. 16.4 Move from the stable ground to aggravating slope instability; unavoidable situation or mismatching of the principles of disaster management against landslides

16.3 Slope Instability and Degradation

Landslides or slope instability deal with natural resources such as soil, water, flora, and fauna. It deals with all natural and man-made stresses from the grass root level upto the slope failure event. Natural slope usually contains more vegetation, grass, soil, water, gases, stones, boulders and exposed rock outcrops. Predominant behavior of a slope is standing without any disturbing of stress history. Stress history of a ground means the ground subjected to various stresses due to climatic and geological deformities. Therefore, ground may or may not contain any form of residual stresses to move soils from hill slopes due to extrinsic variables. Gravitational stress is decreased as an unstable stress due to known history of weathering of soil. Therefore, slopes are usually categorized into three major slope segments, namely crown, middle, and toe as shown in Fig. 16.5. The top part of the slope may have dip or escarpment slope which always tends to destabilize with minimum potential energy. Usually, natural slopes are more likely to be concave and showing these three major slope segments until reaching the flat surface to attain natural stability. The middle slope usually contains secondary deposits of upper most section of the slope. The toe or the lower section of the slope shows long history of stability with low or non of residual stress.

16.4 Observations and Inventories

Observations and inventories are the most important datasets in every natural disaster mitigation and management. The geoscientists are experts of understanding the requirement of inventories of landslides and use such inventories to predict and safeguard the community against re-occurrences of landslides (Rupasinghe et al.

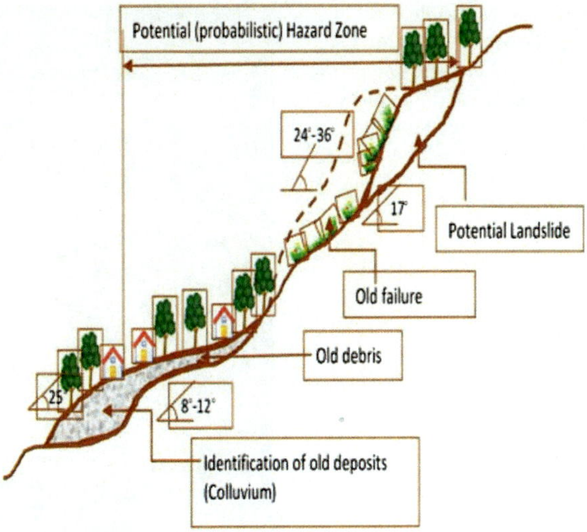

Fig. 16.5 Principles of natural degradation of slopes and demarcating of landslide hazard potential zones

2014). The historic information on landslide events is a fundamental requirement as it provides an insight into the frequency of the phenomena, the types of landslides, the volumes of materials involved, and the damage caused. Therefore, it is required to built-up a mechanism to collect disaster-related databases and studies and disseminate among the community who are living in vulnerable areas. In most cases, communities have forgotten the past events and re-occurrence possibilities. Some are ignorant due to long return period of the disaster event. The historic information on landslides requires data on topography, geological settings, geomorphology, geometry, soil type, soil depth, geotechnical characterization of soil materials, vegetation, hydrology, hydrogeology, land use, and landform (Lakmali et al. 2014).

16.5 Long Travel Landslides

Recently, rain-induced long travel complex form of landslides were recorded in high numbers with casualties from Sri Lanka, Philippines, and Nepal. Rapid landslides mean debris flows, debris avalanches, and rock falls and rock avalanches (Hungr and Evans 2004a). Slow, ductile toppling of rock masses commonly creates large-scale mountain slope deformations (Hungr and Evans 2004a). The occurrence and classification of massive rock slope failure. Felsbau, Vienna, Austria 22:16–23. Such cases suddenly develop into a catastrophic failure which is extremely rapid flow slides consists with soils and rocks/boulders (Hungr et al. 2001). Typical examples are Mulahalkele Landslide (Fig. 16.6), Ragala, Nuwara Eliya district in 1986 and the Aranayake landslide (See Fig. 16.7) in 2016 in Kegolle district, Sri Lanka. The failure behavior of landslide in clay soil and residual soil is very different from above. Some continues to exhibit intermittent, relatively slow deformations with limited mobility.

Fig. 16.6 Mulhalkele Landslide in 1986. Zero victims and damage to the general hospital premises at the Mulahalkele, Nuwara Eliya district in Sri Lanka (NBRO 1988)

It is also noted that toppling or deep weathered rock strata can initiate a brittle catastrophic rockslide with the residual soil under fully saturated conditions (Hungr and Evans 2004a).

Another inventory of devastated and long-travel landslide has been reported on May 17th, 2003, killing seven people and destroying eight houses in Pothupitiya, Kalawana, Sri Lanka. According to the observations, it could be concluded that the fundamental causative factors for occurrence of the Pothupitiya Kanda landslide were continuous excessive precipitation within a short period of the day, existence of the faults, and prominent rectangular joint pattern, deeply weathered thick overburden and the change of land use from forest to tea cultivation. The landslide covered approximately one square kilometers area and occurred as a debris flow, taking two paths along the existing valleys and ultimately causing temporary damming a Delgodaganga river (Dias et al. 2011). A few minutes after failure caused inland flooding below the low ground areas in Ratnapura district. That was the first record in Sri Lankan disaster history on a landslide dam disaster and unexpected number of causalities due to low-land inland flooding (Fig. 16.8).

Fig. 16.7 Landslide event captured more than 127 victims in Aranayake, Kegolle district in Sri Lanka (JICA 2016)

Fig. 16.8 May 17th, 2003, killing seven people and destroying eight houses in Pothupitiya, Kalawana and caused temporary damming at the Delgodaganga River

16.6 Rainfall as a Key Parameter

The slopes generally fail at some point following a high level of rainfall precipitation over a prolonged period of time. It has been indicated that the Naketiya Earthslide which was initiated in the nineteenth decade, and reactivation is continued until to date and recorded as the longest earthslide in Sri Lanka.

Similarly, most of the landslides investigated to date are known to be rain triggered, seismic triggered, or both. Analysis of rainfall records of very large number of Sri Lankan landslides conveys that 24 h rainfall associated with a landslide event was generally 2–23 times higher than the average daily rainfall of that location (Bhandari and Dias 1996). The month in which a particular landslide event fell attracted 2.4–3.5 times the average monthly rainfall, and the annual rainfall in the year of a landslide event was found to be 1.2–1.4 times higher than the annual rainfall average (Bhandari and Dias 1996). Studies on the past occurrence of landslides in the hill districts of Sri Lanka study suggest that if the cumulative rainfall on three consecutive days exceeds 200 mm, and if the rains are found to be continue, then the probability of landslides occurrence should be considered high (Bhandari 1992).

16.7 Viewing Landslide Studies in the Larger Context of Disaster Risk Reduction

The landslide disaster mitigation adoptive policies should drive the agenda of integrating landslide risk reduction with development planning (Rupasinghe et al. 2014). Landslide disaster risk reduction strategy itself should be based on integration of the essential inputs from all related disciplines and sub-disciplines which are critical to understand the interplay of various causative and influencing factors.

It is also important to understand the social acceptability of risk means (Dias et al. 2003) that the society is convinced that landslide risk does not exceed an acceptable risk level when planning future land use. Similarly, residual risk or tolerable risk is a level of risk that a society prepared to live with because there are net benefits in doing so as long as that risk is monitored and controlled, and action is taken in to reduce the risk (EMSA/OP/10/(2013). Once the policy framework is clearly defined for a particular task, the applicable risk minimization standards and guidelines can be implemented accordingly (E-conference 2015–1).

Major criterion that governs on promoting the multi-disciplinary culture of landslide studies which is none other than the landslide professionals. It is essential to create a structure from a large well-branched network of professionals to acts as a tremendous supportive system of sharing information and services among the individuals and groups until information reaches the grass root level (E-conference 2015–3 and 3, 2017-1 2017-2 2017-3, Dias et al. 2017).

16.8 Community-Centered Early Warning Approach as an Essential Tool

People living in the landslide prone areas may be physically, economically, socially, ecologically, politically, and organizationally vulnerable. Though disaster management agencies have taken tremendous efforts to improve preparedness for landslides, certain gaps, shortfalls, or unattended underlying challenges hamper effectiveness of these efforts. People are impressed by the magical power of modern technology but in the context of developing world, it would be inappropriate to overlook the advantages of ease, economy, accessibility, and time-tested benefits of traditional technologies (Dias et al. 2017). It is also the time to think in terms of community-centric early warning systems. Some of the key observation on affected person in a landslide disaster situation is indicated in Fig. 16.9.

Community-centered awareness is a more practical approach which safeguards people (Dias and Wijewardana 2002). There could also be attitudinal and communicational vulnerability. It is therefore imperative to investigate not only the nature of the hazard but also the underlying characteristics of the environment and society that make community prone to risks. Identifying landslide prone areas and vulnerable communities, setting up an awareness raising mechanism, establishing effective early warning systems, developing a shared response plan ensuring timely support for the vulnerable people will minimize loss of lives, injuries, and property damages including some consideration as listed below.

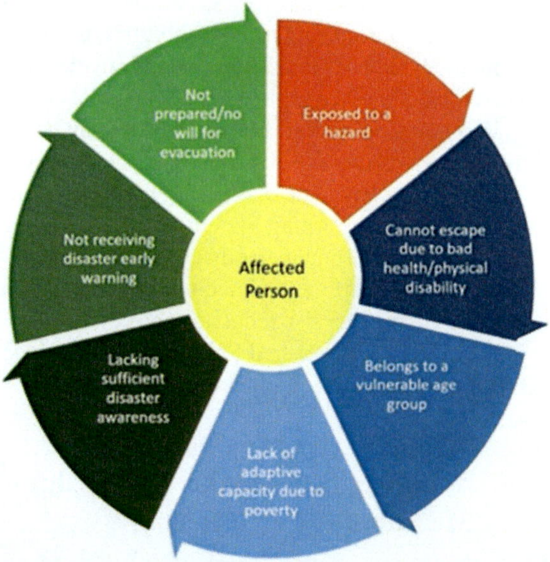

Fig. 16.9 Affected person in a landslide disaster risk

- In most situations, landslide and floods both together create tremendous effects to the vulnerability of ground specially community living in lower catchment of slopes.
- In large landslides, the reoccurrence period usually exceeds more than 100 years and therefore no signatures of old landslides at the time of settlements
- Community can understand the rainfall sensitivity but are unknown about the sensitivity of extent of damage.
- Technology on early warning systems to be filed with timely alarming technology and proper networking facility.
- Probability of an alarming event should be on wide spectrum of data, sound judgments and coupled with experiences.
- Society of not more than 25 years old but much aware the vulnerability issues related to the climate change and ready to accommodate more principals and less negligence.
- Settlers of more than 25 but less than 75 years old; such community usually respect the scientific declarations but will not totally accommodate without having knowledge on same experiences before.
- Settlers of more than 75 years old having less attention on new alarming systems and therefore need more public awareness.
- More advanced setups of early warning technologies will help to safeguard the community and people living in sloping terrains.
- Sending the correct message at the right times to the correct audiences, and with the proper terminology is essential. Reducing vocabulary sophistication, and volume of words is another aspect (World Bosai Forum IDRC 2017).
- Identify a social safety-net mechanism to assist poor and at-risk groups, such as older persons, persons with disabilities, displaced persons, migrants and other populations exposed to disaster risk and affected by disasters (IDRC Report 2017).
- Therefore, it is essential that the communities are empowered to promote simple and economical instrumentation and training to foresee the danger of landslides for themselves (Dias et al. 2017). The community-centered early warning system for mitigating landslide disasters is the development of a highly adaptive monitoring and early warning system (Dias and Dias 2002, 2003), so that the local people and governmental officers are capable of preparing and operating the system. Moreover, the updated schematic diagram for the community-centered approach in disaster risk reduction and emergency planning in landslides is shown in Fig. 16.10 and expected to be able to reproduce the system independently so that it can be applied in other disaster situation as well.

16.9 Acquring Indegenious Knowledge on Slope Stability Measures

The human settlements are more likely recommended in almost flat terrains and maximum slope not exceeding 17 degrees slope to the horizontal. It does not

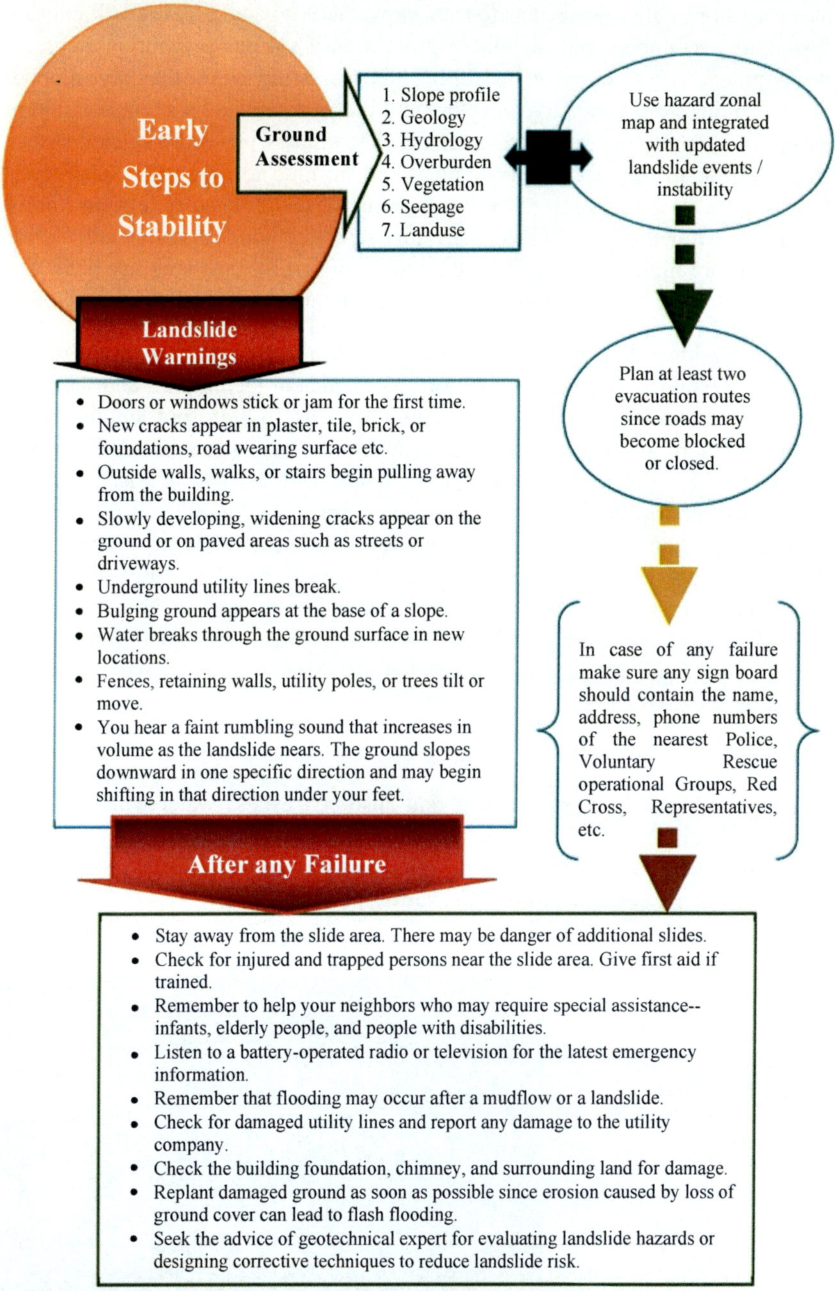

Fig. 16.10 Updated community-centered approach in disaster risk reduction and emergency planning in landslides (Dias et al. TIEMS 2001 and www.fema.gov./library/landslif.htm)

mean up slopes are always unsafe to occupy. Geology and locality will educate community to understand the stable ground with various precaution steps for development with zero risk on instability. It is interesting to note that large number of indigenous slope protective development is characterized by a combination of anti-erosion and soil fertility maintenance techniques pits and ridges on steep slope similar to Fig. 16.11 practicing in old farming techniques in Sri Lanka. Such process was first noticed even in farming of paddy in steep terrains known as "helmalu" method which safeguard fully saturated zone for paddy cultivation. The areas are highly protective by the means of fertility and stability both. The process is lasting from the King's period of the country and continues even today.

Vegetative slopes that had high root density (due to dense vegetation on the surface) are noticed in hill county slope management techniques, and those are less likely to undergo slope failure. This is because a high root length density results in more absorbed water and standing low soil water content which in turn results in an increase in shear strength and decrease soil permeability. It is suggested that root length density and soil water level could be used as indicators of slope stability and possibly could be used to predict future slope failure. Number of locations was observed in road-side bank stability with native plants and standing more than 50–60 years in record. It is interesting to note that annual average annual rainfall in such localities are more than 4000–5000 mm per year and standing without showing unexpected instabilities or extensive erosions as shown in Figs. 16.12, 16.13, 16.14.

Plant roots act in several ways to increase slope stability:

1. Roots bond unstable soil structure and absorbed excess water which leads to stable subsoil,
2. They provide a cover of complex net form structure and speeding over a laterally strong fine root systems close to the surface, and

Fig. 16.11 Paddy fields in Walapana; Indigenous approach of paddy cultivation in Sri Lanka — "Helmalu" method (mountains terrain; https://walapane.com/media/paddy)

Fig. 16.12 Standing of almost straight earth cut in completely weathered rock with vegetative growth

Fig. 16.13 Typical native multi-culture vegetative spices on a saturated rock soil composite roadside; stable cut slope (standing more than 40 years)

Fig. 16.14 Native mono-culture vegetation on overburdened soil standing above the completely weathered rock in Avissawella — Nawalapitiya road

3. In addition, deep rooted trees will provide embedded stems act like a buttress pile or arch-abutment on a slope.

16.10 Conclusion and Recommendation

The development of a policy framework can be done for a diversified extent including promote disaster risk awareness, scientific background of causes and effect, and relationship between the social and professional inputs of landslide disaster management. However, many of the feasible concepts developed for landslide disaster mitigation have not reached the implementation stage successfully due to various reasons (Dias et al. 2017). Thus, each and every individual should build up their own moral strength of acting upon a landslide occurrence with a sensible approach in taking off from such localities and settling down in safer surroundings by focusing on technical and social aspects as shown in Fig. 16.15.

- The landslide risk could not be any longer considered in isolation. The other hazardous events such as flood and earthquakes are also entangled with landslide to a certain extent. It can be further introduced as landslide-triggered earthquake-induced landslides and flood-induced landslides. Since a landslide prone area could also be flood and earthquake prone, it is only by the production of multi-hazard maps that the scientists can connect the universe of landslides with the galaxy of other disasters and encourage holistic approach to hazard investigation and disaster risk reduction including network and strategic approaches as indicated in Fig. 16.16.
- Networking and strategic planning approaches requirements.
- People are well aware about climate issues, climate adaptation principles, or public awareness on climate change adoption issues, etc.
- Early warning systems need to strengthen through new technologies with community-centered early warning system for emergency management of landslides
- Use specially designed community awareness activities considering age groups, educational group, and gender
- Strengthen the implementation of social safety-net mechanisms to assist the poor and at-risk groups, such as older persons, persons with disabilities, displaced

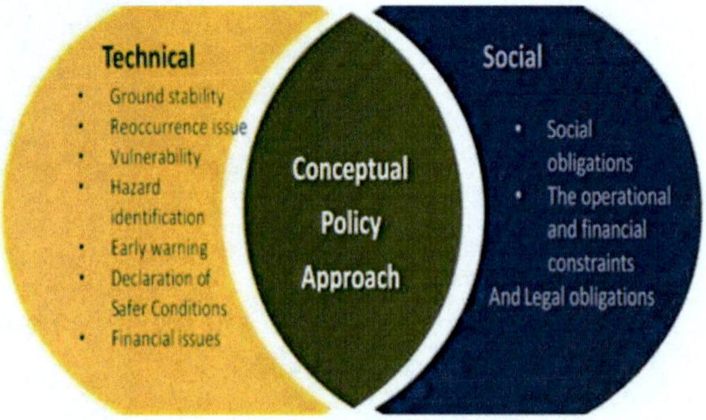

Fig. 16.15 Technical and social issues on conceptual model policy approach in landslide disaster mitigation works

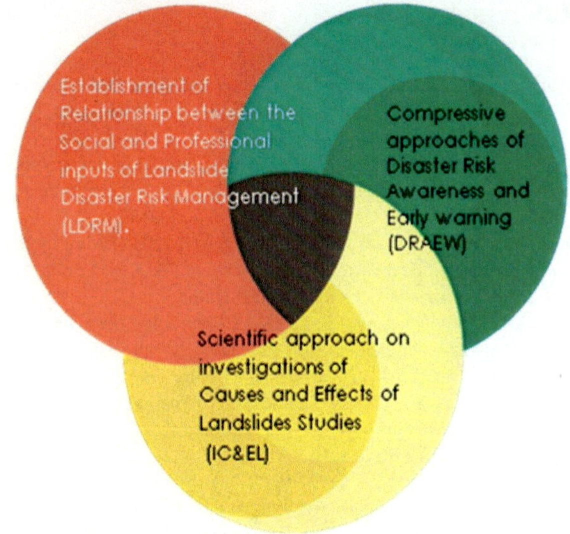

Fig. 16.16 Logical approach of framework to promote multi-disciplinary culture of landslide studies and risk management

persons, migrants and other populations exposed to disaster risk and affected by disasters; (Fact Sheet 2015; www.fema.gov./library/landslif.htm)

- Strengthen the sustainable use and management of ecosystems and implement integrated environmental and natural resource management approaches that incorporate disaster risk reduction.

Such moral strength should be built-up bit by bit until they identify to avoid the vulnerability including cultivation in slope areas, establishing domestic surroundings in mountainous regions and occupying in landslide prone regions for developments (Dias et al. 2017).

Such attempt can only be succeeded by lasting a greater interaction between professionals pursuing landslide studies, and those considering at other types of hazards, in pursuit of the common cause of advancement of knowledge and human safety. The ancient cities in mountain terrain are always standing in lower or the third segment of slope and educate us to understand the engineering and environmental sustainability of the nature.

Acknowledgements This paper forms with integral parts of the landslide research works which have been implemented by the Center for Research and Development, Natural Resources Management and Laboratory Services, and Central Engineering Consultancy Bureau (CECB) of the Ministry of Mahaweli Development and Environment, Sri Lanka. It is published with their permissions. The views expressed in the paper are however those of the authors only. It is our pleasure to acknowledge Ms. Dimuthu Sewwandi for her valuable schematic images and continuous corporation extended for the preparation of this paper. Our grateful thanks are due to the Eng. G. D. A. Piyathilaka, Chairman, and Eng. K. L. S. Sahabandu, General Manager, Central Engineering Consultancy Bureau for the permission and encouragements.

References

Bhandari RK, Dias AAV (1996) Rain triggered slope movement as indicators of landslide dynamics. In: 7th International symposium on landslides, Trondheim, Norway, 17–21 June 1996, Balkema, pp 1515–1520

Bhandari RK (2015) Developing conceptual model policy frameworks to understand causes effects and mitigations of landslide occurrences. In: International E-conference on landslide disaster reduction, Sri Lanka 24th March–24th May 2015

Dias AAV, Rupasinghe N, Gunathilake AAJK (2011) Joint technical emergency operation experiences on landslide disaster mitigation event 2003, Sri Lanka. In: Proceedings of the second world landslide forum—3–7 October 2011, Rome

Dias AAV, Wijewardana PR (2002) Community base participatory model in natural disaster preparedness—landslides. In: Proceedings 9th world emergency management conference, Toronto, Canada, 14–17 May 2002. TIEMS

Dias AAV, Dias SV (2002) Key to environmental assessment model in landslides by observational method of approach. In: International workshop on environmental geomechanics , Ascona, Switzerland, June 30 to July 5

Dias AAV, Dias SV, Siva GVMde (2003) Integrated environmental emergency response stage in natural disaster preparedness—landslides. In: Proceedings 10th world emergency management conference, Nice, France from May, 2003. TIEMS

Dias AAV, Katuwala N, Herath HMJMk, Perera PVIP, Sahabandu KLS, Rupasinghe N (2017) Model policy frameworks standards and guidelines on disaster reduction (WCoE 2014–2017). In: Advancing culture of living with landslides. vol 1. ISRD-ICL Sendai Partnerships 2015–2025, pp 365–374

Dias AAV, Hennayake SK (2014) Role of intervening agencies and officials in emergency risk management of landslides, Sri Lanka. In: Proceeding of the world landslide Forum3 (WLF3), Beijing, China, 2–6 June 2014. vol 4. pp 700–705

Dias AAV, Gunathilaka AAJK (2014) Evaluation of sensitivity of the WAA and SINMAP models for landslide susceptibility risk mapping in Sri Lanka. In: Proceeding of the world landslide Forum3 (WLF3), Beijing, China, 2–6 June 2014. vol 2. Landslide Science for a Safer Geoenvironment, pp 167–173

Dias AAV, Abayakoon SBS, Bhandari RK (2014) Discrete boundary shear strength of a landslide at high rainfall precipitation zone in Sri Lanka. In: Proceeding of the world landslide Forum3 (WLF3), Beijing, China, 2–6 June 2014. vol 1. Landslide Science for a Safer Geoenvironment, pp 101–106

Hungr O, Evans SG (2004a) The occurrence and classification of massive rock slope failure. Felsbau, Vienna, Austria. vol 22. pp 16–23

Hungr O, Evans SG (2004b) Entrainment of debris in rock avalanches an analysis of a long run-out mechanism. Bull Geol Soc Am no. 9/10 116:1240–1252

Hungr O, Evans SG, Bovis M, Hutchinson JN (2001) The Varnes classification of landslide types an update review of the classification of landslides of the flow type. Environ Eng Geosci 7:221–238

EMSA/OP/10 (2013) Risk acceptance criteria and risk based damage stability. Final Report 2013 part 1: risk acceptance criteria, European Maritime Safety Agency Norway

E-conference (2015–1) Developing conceptual model policy frameworks to understand causes, effects and mitigations of landslide occurrences. In: Proceedings of first e-conference on landslide disaster risk reduction 24th March–24th April 2015, WCoE, CECB. https://e-conference.crdcecbsl.lk

E-conference (2015–2) Originate pertinent standards for humanitarian activities in support of effective risk reduction and mitigations on landslide occurrences. In: Proceedings of first e-conference on landslide disaster risk reduction 24th March–24th April 2015, WCoE, CECB. https://e-conference.crdcecbsl.lk

E-conference (2015–3) Implementation of applicable guidelines/teaching tools to establish essential synergies in landslide disaster phenomena. In: Proceedings of first e-conference on landslide disaster risk reduction 24th March–24th April 2015, WCoE, CECB. https://e-conference.crdcecbsl.lk

E-conference (2017–1) Youth empowerment in landslide disaster risk reduction through knowledge sharing. In: Proceeding of 2nd e-conference on landslide disaster risk reduction 30th April–30th May 2017, WCoE, CECB. https://e-conference.crdcecbsl.lk

E-conference (2017–2) Disaster resilient housing, building, land management and agricultural development. In: Proceeding of 2nd e-conference on landslide disaster risk reduction, 30th April–30th May, 2017, WCoE, CECB. https://e-conference.crdcecbsl.lk

E-conference (2017–3) Indigenous knowledge in landslide disaster management. In: Proceeding of 2nd e-conference on landslide disaster risk reduction, 30th April–30th May 2017, WCoE, CECB. https://e-conference.crdcecbsl.lk

Fact Sheet (2015) Landslides and mudflows, www.fema.gov./library/landslif.htm

Herath HM, Janaki MK, Jayasooriya JADNA, Dias AAV (2016) Geological stability of overhanging rock slopes. In: Proceedings of 2016 IPL symposium, UNESCO, Paris, November 17–18

Hutchinson JN (1988) Morphological and geotechnical parameters of landslides in relation to geology and hydrogeology. In: Bonnard C (ed) Landslides. Proceedings 5th international conference on landslides. lausanne. vol 1. pp 3–35

IDRC Report (2017) World Bosai Forum IDRC 2017, Sendai, Japan from November 25–28

JICA Survey Team (2016) Aerial Survey report on inundation damages and sediment disaster. https://www.jica.go.jp/srilanka/english/office/topics/c8h0vm00006ufwhl-att/160615.pdf

Landslide Dynamics Vol 1 (2018) ISDR-ICL Landslide interactive teaching tools. Fundamentals mapping and monitoring, vol 1. Springer Nature

Landslide Dynamics Vol 2 (2018): ISDR-ICL Landslide interactive teaching tools. Testing risk management and country practices, vol 2. Springer Nature

Lakmali MADC Herath HMJMK Kodagoda SSI Andree SNL Dias AAV 2014 Database designing for the road affected landslides and quantitative evaluation of landslide risk. In: Proceedings of the world landslide forum. vol 3. Beijing, China

Mallawarachchi MASN, Ekanayake EMTM, Kodagoda SSI, Dias AAV (2014) Comparison of soil modulus E50 of residual soil slope failures in two different rainfall zones. In: Proceeding of the world landslide Forum3 (WLF3), Beijing, China, 2–6 June 2014. vol 1. Landslide Science for a Safer Geoenvironment, pp 135–141

McDonald Rk (2003) In: Introduction to natural and man-made disasters and their effects on buildings

NBRO (1986 and 1988) Study of Landslide Hazard in Sri Lanka

Rupasinghe N, Dias AAV, Hennayake SK (2014) Role of intervening agencies and officials in emergency risk management of landslides, Sri Lanka. In: Proceeding of the world landslide Forum3 (WLF3), Beijing, China, 2–6 June 2014. vol 4. pp 700–705

Chapter 17
Disaster Resilient Infrastructure

Bhumika Gupta and Salil K. Sen

Abstract Inducting the sustainable infrastructure rating systems enable assessment of the degree to Sustainable Development Goals (SDGs) compliance (Diaz-Sarachagaa et al. 2016). Day-to-day revenue and profit pressures relegate disaster preparedness to lower priorities. Policy-makers, processors, and community are taking serious note of disaster risk management targets and indicators. (Mitchell et al. 2013). Integrated assessments put in parenthesis disaster-proofing, technology innovation, SDG compliance and overarchingly, good governance (Lotze–Campen 2015) Integrated models: integrated assessment). Infrastructure sustainability assessment commenced with societal, environmental and economic costs, and value added spanning their life cycles. This was boosted to the next level with disaster preparedness and resilience charted by Hyogo and Sendai frameworks that focused on reduction of global disaster damage to critical infrastructure, access to multi-hazard early warning, strengthen resilience of aging and new critical infrastructure, spanning, water, transportation, telecommunications, education, health care, and sanitation (UNISDR 2015).

B. Gupta · S. K. Sen (✉)
IIM, Shillong, India
e-mail: salil.sen@gmail.com; salil.sen@imt-bs.eu

B. Gupta
e-mail: Bhumika.gupta@imt-bs.eu

SUMAS, Geneva, Switzerland

IMT BS, Evry, France

RMUTT, Bangkok, Thailand

RMUTT, Singapore (Intermittent), Singapore

RMUTT, Shillong, India

RMUTT, Geneva, Switzerland

RMUTT, Evry, France

RMUTT, Bangkok, Thailand

H. Tatano and A. Collins (eds.), *Proceedings of the 3rd Global Summit of Research Institutes for Disaster Risk Reduction*, Disaster and Risk Research: GADRI Book Series,
https://doi.org/10.1007/978-981-15-8662-0_17

Keywords Disaster resilience · Infrastructure immunity · Metabolism of sustainable growth

17.1 Financing for Disaster Resilient Infrastructure

The sector-specific allocation of funding has a significant correlation with disaster development (Table 17.1).

Disaster Risk Reduction and Management (DRR & M) had potential to provide *immunity* to the *metabolism* of sustainable growth pivoted disaster knowledge access, safety nets and resilience to disaster-led risks, and learning design solutions (Wamsler 2009).

17.1.1 Key Objectives

(i) Enhance the effectiveness on disaster preparedness
(ii) Link the vast repositories of wisdom at locales in ASEAN/Asia Pacific and beyond
(iii) Create community of practice on Disaster Risk Reduction and Management
(iv) Road map for systematic disaster mitigation–adaptation–rehabilitation–sharing steps
(v) Incorporate DRRM as an integral part of development management with private–public–community linked architecture that breaks away from *silo* like existence of Disaster Management.

17.2 Literature Review

Disaster resilience disaster risk reduction and management aims to foster content and empathy, skills for proactivity, creativity and innovativeness, initiative and self-direction, social and cross-cultural adaptivity (Gosper and Ifenthaler 2014). Preparedness to cope with the progression of vulnerability to disasters by embedding resilience

Table 17.1 Sector-specific allocations (approx.)

Sector clusters	Allocations (%)
Water supply, sanitation, and waste management	Over 30
Transport and communications	About 28
Multi-sector	About 10
Energy, health, education	About 20

to cope. Quality, knowledge, and innovation practices in DRR & M are tabulated (Table 17.2).

In East Asia and Pacific, weighted mainly by PRC, infrastructure investments, estimated by bottom-up methodology, surpasses demand, determined by top-down process. This anomaly necessitates review of methodology to create the infrastructure financing bankable to private sector.

17.3 Issue Map of Disaster Risk Reduction and Management with Respect to Infrastructure

Related to disaster risk reduction and management are reliability of energy–water–waste assurance at pre-stages, during post-stages of disasters. There is a broad-band of issues, spanning infrastructure, agriculture, education among others with soft issues on governance, empathy, preparedness that encapsulated the scope of DRR & M (Table 17.3).

17.4 Community of Practice Approach for DRR & M

Energy infrastructure financing to address the deficit of about 7 to 8% (Table 17.4). In a related study, infrastructure investment demand in Emerging Markets and Developing Economies (EMDE) for in top-down methodology is USD 356 bi while bottom-up need is USD 212 bi (Ruiz-Nunez and Wei 2015) (Table 17.5).

References: RSDD sector and thematic priorities; long-term strategic framework 2000–2015; supporting finance at ADB; knowledge management directions and action plan 2013–2015, supporting finance + +

Table 17.2 Quality, knowledge, and innovation practices in DRR & M

Disaster rebuilding clusters	Interfaces with other development management fields of study
Agriculture, rural development, and food security:	Encompass sector reviews, financing practices, and instruments and support for agricultural productivity and improvements
Climate change mitigation and adaptation	Include mitigation and adaptation strategies, climate change implementation plans, climate change fund operations, external consultations
Disaster risk management	Comprise disaster risk management, disaster and emergency assistance, risk and vulnerability assessment, policy reviews, good practices, external consultations

(continued)

Table 17.2 (continued)

Disaster rebuilding clusters	Interfaces with other development management fields of study
Education	Coordinating education sector policy and strategies, monitoring the education sector plan of action, advisory support to education sector operations, good practices, partnerships and innovation (UNCRD 1996), extreme event analysis (Smith 1996)
Environmental services	Develop proactive environmental programs, mainstreamed environmental considerations
Environmental and social safeguards	Introducing safeguards policy compliance, policy reviews, advisory services
Poverty reduction, social development, gender and development, and health	Poverty reduction and social development: millennium development goals, regional poverty networks, country poverty analysis, impact assessments, community-driven development, good practices, social protection, promoting labor standards Gender and Dev: monitoring gender and development policy, operational advisory support, good practices, partnerships, participatory approaches Health: fostering better health outcomes, helping monitor health impacts of related ADB projects, promoting health as a regional public good (HIV and AIDS, communicable diseases), sharing knowledge and experiences
Public management, governance, capacity development	Institutional diagnosis, governance and capacity development, public financial management, service delivery mechanisms, anti-corruption initiatives, consultations with civil society and non-government organizations
Knowledge management	Weaving knowledge management in operations, communities of practice, knowledge hubs, learning and development, technology

Table 17.3 Issue map of DRR & M

Key stakeholders	Level of engagement
Policy, public–private partnerships	Board, management, staff, country, and sector teams

(continued)

Table 17.3 (continued)

Key stakeholders	Level of engagement
Local stakeholders	Civil society, non-governmental organizations
Umbrella organizations	Development agencies, academia and research institutions
Sector assessments and road maps	Agricultural and natural resources sector, transport, information and communication technology
Country partnerships strategy assessments	Energy, finance, education
Governance risk assessments	Health and social protection
Economic analysis and financial retrospectives	Industry and trade, public sector management
Safeguards and mitigation	Climate change

[a]Post-completion sustainability assessment study has a database of 548 sustainability ratings criteria.

Table 17.4 The community of practice (CoP) approach disaster risk reduction and management course content

Disaster preparedness linked CoP keywords:	Key characteristics:
Rationale for communities of practice (CoPs) on DRR & M:	CoPs are groups of like-minded: shares know-how; creates thematic–sectoral "linked" domains of knowledge; continual anticipate—know—build—adopt cycles
Finance for disaster risk reduction	Models of financing rehab, preparedness, private–public partnerships
Education/capacity building for disaster preparedness	Four key attributes (i) empowerment of local inhabitants, (ii) capacity building of local businesses, (iii) livelihood continuity, (iv) professional/vocational training
Trans-boundary collaborations/consortiums to cope with disasters	Key CoP needs: (i) sharing knowledge that conforms to local typicality, (ii) amplify little-understood, little known ideas on disaster prevention *wisdom*, (iii) provides resources to enable carrying out idea, (iv) convenes different groups may be online, (v) continuity issues of community of practice
Disaster database	Database needs filtering on disaster relevance; necessary amalgamations, knowledge management action plans For instance, ADB community of practice has 19 committees, 9 networks with respect to sectors, 10 themes
Four pillars of disaster preparedness	(a) knowledge to action focus, (b) empowering, (c)trans-boundary partnerships, (d) enhancing multi-livelihood learning and re-skilling in face of disasters

Table 17.5 Key output measures of DRR & M effectiveness

Catalyzing investment	Monitoring quality and compliance	Promoting regional cooperation and integration
Strengthening inclusiveness	Preventing corruption	Managing the environment
Improving governance	Assess and monitor risk	

References

Aitsi-Selmi A, Egawa S, Sasaki H, Wannous C, Murray V (2015) The Sendai framework for disaster risk reduction: renewing the global commitment to people's resilience, health, and well-being. Int J Disaster Risk Sci 6(2):164–176

Asian Development Bank (2014) Operational plan for integrated disaster risk management 2014–2020. Manila: Asian Development Bank, April 2014. Available at: https://www.adb.org/sites/default/files/institutional-document/42764/files/integrated-disaster-risk-management-operational-plan.pdf

Diaz-Sarachagaa J, Jato-Espinoa D, Alsulamib B, Castro-Fresnoa D (2016) Evaluation of existing sustainable infrastructure rating systems for their application in developing countries. Ecol Indi 71:491–502

Gosper M, Ifenthaler D (2014) Curriculum models for the 21st Century: using learning technologies in higher education, Springer

Greve C, Hodge G (2013) Rethinking public–private partnerships: strategies for turbulent times, Routledge

Kesel J, Sedlak V (2014) Global climate change and security threats. Eur Sci J 10(20)

Lotze−Campen H (2015) Integrated models: integrated assessment

Maurrasse D (2013) Strategic public private partnerships: innovation and development, Edward Elgar Publishing

Milosevic DZ (2003) Project management tool-box: tools and techniques for the practicing project manager, John Wiley & Sons

Morelli S, Segoni S, Manzo G, Ermini L, Catani F (2012) Urban planning, flood risk and public policy: the case of the Arno River, Firenze, Italy. Appl Geogr 34:205–218

Mitchell T, Jones L, Lovell E, Comba E (2013) Disaster risk management inpost-2015 development goals: potential targets and indicators. Overseas Development Institute, London, pp 71–85

Prasad N, Ranghieri F, Shah F, Trohanis Z, Kessler E, Sinha R (2008) Climate resilient cities: a primer on reducing vulnerabilities to disasters. The World Bank

Ruiz-Nuñez F, Wei Z (2015) Infrastructure investment demands in emerging markets and developing economies. The World Bank

Smith K (1996) In: Environmental hazards: assessing risk and reducing disaster

UNCRD (1996) Innovative urban community development and disaster management proceedings

UNISDR C (2015) The human cost of natural disasters: a global perspective

Wamsler C (2009) Urban risk reduction and adaptation. Verlag Dr, Muller

Weiss J, Potts D (2012) In: Current issues in project analysis for development, Edward Elgar

Wisner B, Gaillard J, Kelman I (2012) Handbook of hazards and disaster risk reduction, Routledge

World Bank (2014) World development report

Chapter 18
Drought Disaster in the Central Highlands of Vietnam: Relationship Between Land-Use Change and Drought Impact

Nguyen Thi Thu Ha, Mai Trong Nhuan, Dinh Ngo-Thi, and Nguyen Thien Phuong Thao

Abstract Over the past decades, drought becomes a global hazard with nearly half of the world's countries suffered from drought, and it is arguably the biggest single threat from climate change. However, human activities which are reflected through land-use conversion are also inducing drought impact. This study aims at quantifying the effect of changes in land use to drought impact by the case of drought in the Central Highland of Vietnam. In this study, drought impact was modeled using Normal Different Drought Index (NDDI) retrieved from Landsat data, and it was multivariate regression analyzed with local statistical data on land-use pattern areas. Result confirms the dependency of drought impact on land-use change, whereas the reduction of local forest land and excessively expansion of industrial cropland are two main factors those lead to the increase of total drought-affected area in the highland (Pearson correlation coefficient, $R = 1.00$ and 0.98, respectively). Modification of agricultural crop plan, particularly modification of cultivated mechanism and crop structure, is an adaptation strategy to drought in the highland.

Keywords Dought disaster · The Central Highlands of Vietnam · NDDI · Crop loss · Land-use · Landsat

N. T. T. Ha (✉) · M. T. Nhuan
Faculty of Geology, VNU University of Science, Vietnam National University, Hanoi, 334 Nguyen Trai, Hanoi, Vietnam
e-mail: hantt_kdc@vnu.edu.vn; thuhageo@gmail.com

M. T. Nhuan
e-mail: nhuanmt@vnu.edu.vn

D. Ngo-Thi · N. T. P. Thao
Sea and Island Research Center, VNU University of Science, Vietnam National University, Hanoi, 334 Nguyen Trai, Hanoi, Vietnam

H. Tatano and A. Collins (eds.), *Proceedings of the 3rd Global Summit of Research Institutes for Disaster Risk Reduction*, Disaster and Risk Research: GADRI Book Series,
https://doi.org/10.1007/978-981-15-8662-0_18

18.1 Introduction

Drought is one of the most widespread and prolonged environmental hazards linked to climate change, especially in less-developed countries with economies that are highly dependent on agriculture such as Vietnam. In recent decades, drought is considered as a severe disaster causing negative impacts on the environment, ecosystems, as well as the economy of the Central Highlands in Vietnam (2016; Huy et al. 2016; Ha et al. 2016; Quyen et al. 2016). As for 2015, the Vietnamese Government has provided 5221 tons of food and allocated 1008 billion VND (~45 million USD) worth of relief and disaster support services for people in the Central Highland's drought-affected regions. According to drought risk records, light drought often reduces 20–30% crop-yields, while severe drought reduces approximately 50% crop-yields, and extremely severe drought may lead to a complete loss of crops (Quyen et al. 2016). In the highland, rice and coffee are the most sensitive crops, while ethnic minorities and small householders are the most vulnerable groups to the impact of drought (CGIAR Research Centers in Southeast Asia 2016). Confronted with increasing drought stress and water scarcity in the region, adaptation to droughts in a changing climate has been a key scientific focus.

According to the United Nations' Millennium Ecosystem Assessment, modification of land use by several anthropogenic activities affects social-ecological systems and therefore, loss of biodiversity (Sarukhán and Whyte 2005). Over time, changes in land use and land degradation can affect the magnitude and frequency of droughts (Sentis 2003). Adjustments in land-use patterns are effective adaptation strategy to drought risk (Lei et al. 2014). National statistical data of Vietnam during 1995–2015 indicated a significant conversion of land use in the highland: from forest land into industrial cropland, particularly coffee cropland (Anh 2015). Therefore, quantifying the relationship between land-use change and drought impact may help mitigate drought risk in the highland by means of future reasonable adjustments in land use.

This study aims at quantifying the relationship between drought disaster impact and change in local land use. In the study, drought disaster impact has been mapped and assessed using Normal Difference Drought Index (NDDI) retrieved from multi-generational Landsat images acquired in Marches (common last month of the local dry season) of years those droughts happened severely: 1993, 1998, 2005, 2015, and 2016. National statistical data on local land-use factors were empirically analyzed with obtained NDDI to quantify the relationship. Multivariate regression of dependency between NDDI and local land-use data on areas of local residential land, forest land, industrial cropland, rice cultivated land, etc. help identify driven factors leading to increasing of drought magnitude and drought-affected region expansion for future adjustment.

18.2 Materials and Methods

18.2.1 Study Area

The Central Highlands is one of eight agro-ecological regions of Vietnam (Fig. 18.1). The region consists of various plateaus surrounded by mountain ranges. The elevations of plateaus range from 500–1500 m above sea level. The Central Highlands has a total land area of 5,454,500 ha (17% of the national area), covering five provinces: Kon Tum, Gia Lai, Dak Lak, Dak Nong, and Lam Dong.

The Central Highlands of Vietnam is well-known as an area of industrial crops with the average GDP growth rate in a period from 2001 until now is 11.9% per year. Because of locating in the upper basin of three large rivers (Sesan River, Serepok River, Dong Nai River), the highland has a great hydro-power reserve where 1560 reservoirs were constructed to provide about 60% of regional irrigation needs (Thanh 2015) and more than 30% of national hydroelectricity potential.

Drought occurs in the highland in yearly dry season, but severely impacted in years El Nino phenomenon dominated such as 1993, 1998, 2002, 2005, 2010, 2015, and 2016 (Ha et al. 2016) with total impact area reach to 40% of the entire land area.

18.2.2 Image Processing Method

Landsat satellites acquire images over the Central Highland of Vietnam from 2:40 to 3:12 GMT (corresponding 9:40 to 10:12 local time) every 16 days following path 124 rows 50, 51, 52 and path 125 row 50. 24 Landsat scenes acquired in March of the years 1993, 1998, 2005, 2010, 2015, and 2016 were used in this study.

With the exception of cloud-masking, all preprocessing of the Landsat images, including radiometric calibration and atmospheric correction, was completed using ENVI 5.3 image processing software. Surface reflectance data from Landsat images were then calculated Normal Difference Vegetation Index (NDVI) and Normal Difference Water Index NDWI based on Tucker (1979) and Gao's methods (1996). NDDI then was calculated using a method proposed by Gu et al. (2007) which defined by the difference between NDVI and NDWI at each image pixel. Accordingly, areas with the NDDI value less than 0.1 are non-drought area; areas with NDDI > 0.1 are under drought-affected region, whereas areas with NDDI > 0.3 are severely impacted by drought.

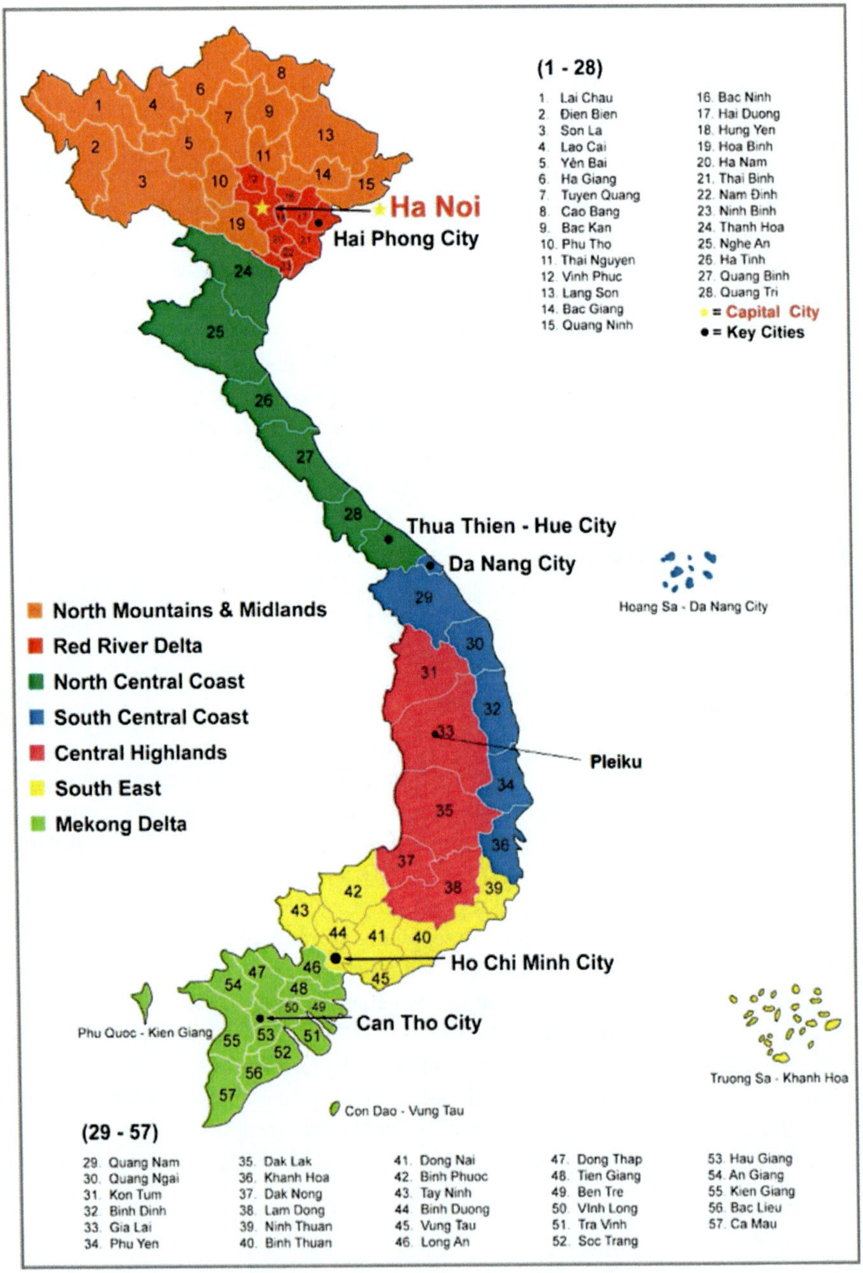

Fig. 18.1 Location of the Central Highlands of Vietnam (pink area) modified from the simple administrative map of Vietnam

18.3 Results and Discussion

18.3.1 Changes in Land-Use and Drought Impact

Resultant NDDI mapping from Landsat data in Marches 1993, 1998, 2005, 2010, 2015, 2016 (Fig. 18.2) showed the most severe drought recorded in March 2005 with total area under severe drought level (NDDI > 0.3) reached to 1,170 thousand hectares corresponding to 21.4% of the highlands' total land area. Drought-affected areas (NDDI > 0.1) have increased dramatically from 34% (in 1993) to 45% of the highlands' total land area (in 2015). The largest drought-affected area was recorded in dry season 2015 with 2486 thousand hectares, corresponding to 45.5% of the highlands' total land area and being larger than itself recorded in 1993 more than 638 thousand hectares although total area under severe drought level was approximately 20% of the highlands' total land area similarly to 1993′s record (Fig. 18.2).

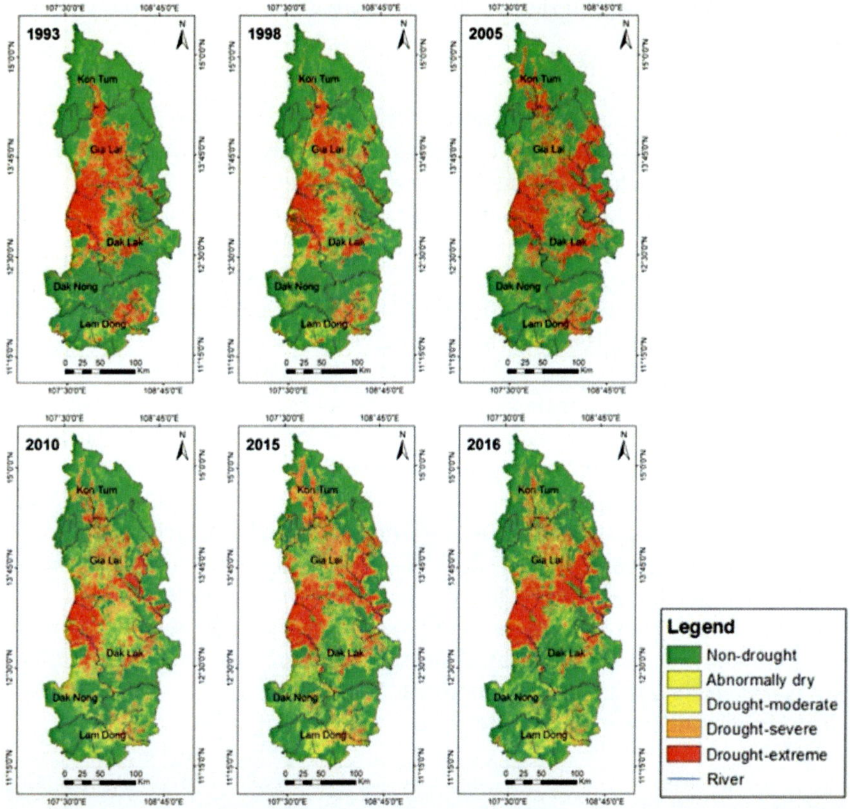

Fig. 18.2 Maps of drought in the central highland of Vietnam in marches of 1993, 1998, 2005, 2010, 2015, 2016

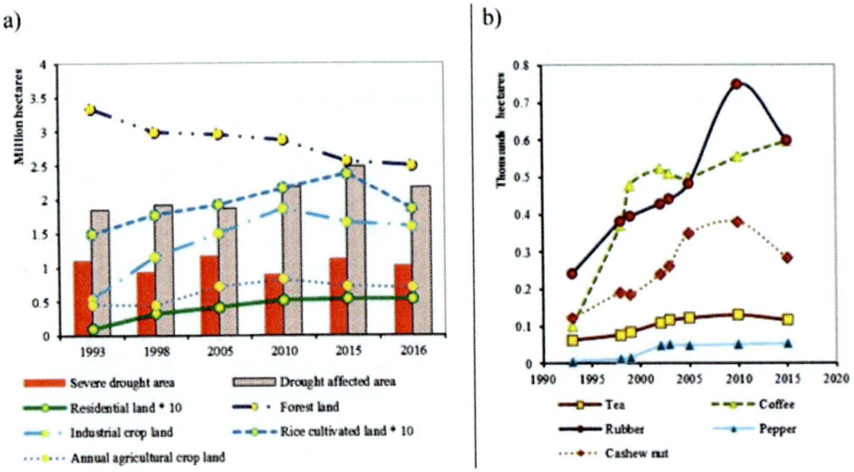

Fig. 18.3 **a** Changes in land-use patterns and drought impact areas, and **b** change in industrial cropland patterns in the Central Highland of Vietnam (*Source* Vietnam General Statistics Office 2016)

According to data from General Statistic Office of Vietnam, during last 25 years, land use in the Central Highlands of Vietnam is changed following the trend of conversion from the natural ecosystem (forest land) into human-made ecosystems (residential land and agricultural lands). Forest land has been significantly decreasing over the last three decades, particularly decreasing of forest bioenergy from 50% of rich bioenergy forest land in 1990 into 22% in 2017 (MARD 2017). Contrast to this trend, residential land and agricultural lands such as industrial cropland, rice cultivated land, and annual cropland have been expanding rapidly for the last 25 years (Fig. 18.3a). Among these land uses, industrial cropland got the highest increasing trend with an *R*-squared coefficient of trend line reaches to approximately 0.65.

Five industrial plants have been mainly cultivated in the Central Highlands of Vietnam are tea, coffee, rubber, pepper, and cashew. According to statistical data, lands used for cultivating these plants have been increasing dramatically in recent years, particularly land used for cultivating coffee with an R-squared coefficient of trend line reaches to approximately 0.68 (Fig. 18.3b). Loss of forest land means loss of water-retention potential, while the increase of industrial cropland and other agricultural lands refers to the increase of water demand in the region in the dry season. Besides, increase in residential land means an increase of impervious surface such as concrete roads and pavements which prevents water exchange between air and soil environment that may lead to more severely impact of drought.

18.3.2 Relationship of Land-Use Change and Drought Impact

Table 18.1 showed the Pearson correlation coefficients (R) of several local climate and land-use factors versus areas under different drought-level impacts. Accordingly, both areas under severe drought impact and total drought-affected area have a negatively strong correlation with dry season average rainfall ($R = -0.86$ and -0.75, respectively) that confirm the most dependency of drought on a reduction of seasonal rainfall than other climate factors such as temperature and length of the dry season. If the area under severe drought impact has no correlation to land-use factors, then the total drought-affected area has strong correlations to the increase of residential land and industrial cropland ($R = 0.70$ and 0.78) which similar to the area under moderate drought impact. These correlations may imply that the expansion of residential land and industrial cropland is the root cause of the expansion of the total drought-affected area.

Table 18.2 showed the Pearson's correlation coefficients between differences in changes of drought-impacted areas versus changes in the land-use pattern. Accordingly, the change in area under severe drought impact was strongly depended on a change in areas of annual cropland and residential land, but the total drought-affected area was strongly correlated to the reduction of forest land ($R = -1.00$) and increasing of industrial cropland ($R = 0.94$).

To check the dependency of the total drought-affected area on local industrial cropland, five common crop types in the highland, including cashew, pepper, coffee, tea, rubber were regression analyzed using their areal data series over time, in 1993, 1998, 2005, 2010, 2015, 2016. Result in Table 18.3 showed the highest dependency of total drought-affected area on coffee cropland, then pepper and cashew croplands.

Because area under severe drought impact has been not changed much during last three decades and mostly depended on seasonal rainfall which human can hardly adjust, the expansion of total drought affected should be focused to mitigate drought impact. Therefore, the reduction of forest land and the increase of industrial cropland trends in the highland should be paid attention in future land-use planning. Among five industrial crop types in the highland, excessively expansion coffee cropland most induces the drought impact over time. This assumption is in accordance with a practical requirement in coffee tree cultivation mechanism: After harvested in November, the coffee tree should be leaves and branches cut for the new generation in spring. And, during spring (corresponding to the dry season in the highland), coffee tree needs more and more water to grow new leaves. This characteristic in coffee cultivation mechanism makes the highland into the severe deficiency of water due to higher water demand for the large area of coffee cropland in a dry season.

Table 18.1 Pearson's correlation coefficient ($N = 6$) between drought areas versus local land use and climate factors over time (1993–2016)

	Rainfall (mm)	Temperature (°C)	Length of season (months)	Residential land (thous.ha)	Forest land (thous.ha)	Industrial cropland (thous.ha)	Annual cropland (thous,ha)	Rice cultivated land (thous.ha)
Severe drought area (thous.ha)	**−0.86**	−0.45	0.12	0.26	−0.27	0.37	0.33	0.14
Moderate drought area (thous.ha)	−0.32	−0.05	−0.32	**0.80**	−0.66	**0.78**	0.62	0.73
Total drought-affected area (thous.ha)	**−0.75**	−0.31	−0.14	**0.70**	−0.62	**0.76**	0.63	0.58

Table 18.2 Pearson's correlation coefficient ($N = 6$) between difference in drought impact areas versus local difference in land-use factors over time (1993–2016)

	DIFF (Residential land,1)	DIFF (Forest land,1)	DIFF (Industrial cropland,1)	DIFF (Annual cropland,1)	DIFF (Rice cultivated land,1)
DIFF (Severe drought area,1)	**0.95**	−0.13	0.47	**0.98**	0.71
DIFF (Total drought-affected area,1)	0.42	**−1.00**	**0.94**	−0.09	0.78
DIFF (Moderate drought area,1)	0.25	**0.98**	**0.86**	0.27	0.66

Table 18.3 Dependency analysis result of total drought-affected areas on local industrial crop types over time (1993–2016)

	R^2	Const	Beta (dependency coefficient)				
			cashew	pepper	coffee	tea	rubber
Total drought-affected area $= f$(cashew, pepper, coffee, tea, rubber)	1.00	528.93	**2.66**	**2.67**	**5.29**	0.00	0.00

18.4 Conclusion

Quantifying the effect of local land-use change and drought impact provides meaningful for future reasonable adjustments in land-use patterns, and it could be a preferred approach in drought risk adaptation. Our findings suggest that reduction of forest land is a key considered issue in the Central Highlands of Vietnam that induces drought disaster. Additionally, an increase of industrial cropland, particularly increase of coffee cropland should be considered as one of the factors that leads to expansion of drought impact in the high land. Modification of agricultural crop plan, particularly modification of cultivated mechanism and crop structure is an adaptation strategy to drought that helps to reduce the physical vulnerability of agriculture to drought (i.e., to reduce drought loss), but also to mitigate the "social vulnerability" of farmers (that is, to improve the farmers' livelihoods).

References

Anh VT (2015) Some issues regarding state administration innovation for land in the central highlands. Vietnam J Sci Technol Eng 10:14–16 (in Vietnamese). Online at https://www.vjol.info/index.php/khcn/article/view/29991/25547+&cd=1&hl=vi&ct=clnk&gl=vn.

CGIAR Research Centers in Southeast Asia (2016) The drought crisis in the Central Highlands of Vietnam Assessment Report. Tum K, Lai G, Lak D Vietnam 2016. Assessed online PDF on 17 Nov 2016

Gao BC (1996) NDWI—A normalized difference water index for remote sensing of vegetation liquid water from space. Remote Sens Environ 58:257–266

Gu Y, Brown J, F Verdin JP Wardlow B (2007) A five-year analysis of MODIS NDVI and NDWI for grassland drought assessment over the central great plains of the United States. Geophys Res Lett 34(L06407):2007

Ha NTT, Nhuan MT, Canh BD, Thao NTP (2016) Mapping droughts over the central highland of Vietnam in El Nino years using Landsat imageries. VNU J Sci 3(1S):1–3

Huy BQ, Kien TT, Hung AQm Long VH Giang NV (2016) Application of multi-temporal satellite images in rapidly assessing drought in the central highland and south central regions. Technical report of DMC (in Vietnamese), Hanoi, p 2016

Lei Y, Wang J, Yue Y, Yin Y, Sheng Z (2014) How adjustments in land use patterns contribute to drought risk adaptation in a changing climate—a case study in China. Land Use Policy 36(2014):577–584

MARD (2017) Status of Forest resource in the central highland of Vietnam. In: Conference on resolutions for sustainable restore of the central highland region towards climate change adaptation in 2016–2010, Dak Lak 20 June 2017 (in Vietnamese)

Quyen NH, Kham DV, Tam T.T (2016) Application of remote sensing in monitoring drought in the central highland of Vietnam. In: National scientific conference on climatology—hydrology—environment and climate change, Hanoi, (in Vietnamese)

Sarukhán J, Whyte A (2005) Millennium ecosystem assessment: ecosystems and human well-being: synthesis. Island Presss, Washington DC

Sentis IP (2003) Hydrological approach for assessing desertification process in the Mediterranean region. In: Desertification in the Mediterranean region: a security issue: proceedings of the NATO Mediterranean dialogue workshop, held in Valencia, Spain, 2–5 December 2003, pp 579–599

Thanh DH (2015) Final report of project "Studying and proposing solutions to enhance capacity of small and medium-sized reservoirs to meet the water demand for production and domestic use and sustainably develop water resources in Tay Nguyen". Project's code: TN3/T30.

Tucker CJ (1979) Red and photographic infrared linear combinations for monitoring vegetation. Remote Sens Environ 8:127–150

Vietnam General Statistics Office (2016) Vietnam National Statistical Year Books of 1990, 2000, 2005, 2010, 2014, 2015